# Lecture Notes in Computer Science     1016

Edited by G. Goos, J. Hartmanis and J. van Leeuwen

Advisory Board:  W. Brauer   D. Gries   J. Stoer

T0239190

**Springer**
*Berlin*
*Heidelberg*
*New York*
*Barcelona*
*Budapest*
*Hong Kong*
*London*
*Milan*
*Paris*
*Santa Clara*
*Singapore*
*Tokyo*

Roberto Cipolla

# Active Visual Inference
# of Surface Shape

Springer

Series Editors

Gerhard Goos
Universität Karlsruhe
Vincenz-Priessnitz-Straße 3, D-76128 Karlsruhe, Germany

Juris Hartmanis
Department of Computer Science, Cornell University
4130 Upson Hall, Ithaca, NY 14853, USA

Jan van Leeuwen
Department of Computer Science,Utrecht University
Padualaan 14, 3584 CH Utrecht, The Netherlands

Author

Roberto Cipolla
Department of Engineering, University of Cambridge
Trumpington Street, CB2 1PZ Cambridge, UK

Cataloging-in-Publication data applied for

Die Deutsche Bibliothek - CIP-Einheitsaufnahme

**Cipolla, Roberto:**
Active visual inference of surface shape / Roberto Cipolla. -
Berlin ; Heidelberg ; New York ; Barcelona ; Budapest ; Hong
Kong ; London ; Milan ; Paris ; Santa Clara ; Singapore ;
Tokyo : Springer, 1995
  (Lecture notes in computer science ; 1016)
  ISBN 3-540-60642-4
NE: GT

CR Subject Classification (1991): I.4, I.2.9, I.3.5, I.5.4

Cover Illustration: Newton after William Blake
               by Sir Eduardo Paolozzi (1992)

ISBN 3-540-60642-4 Springer-Verlag Berlin Heidelberg New York

© Springer-Verlag Berlin Heidelberg 1996
Printed in Germany

Typesetting: Camera-ready by author
SPIN 10486004   06/3142 – 5 4 3 2 1 0    Printed on acid-free paper

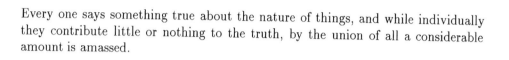

Every one says something true about the nature of things, and while individually they contribute little or nothing to the truth, by the union of all a considerable amount is amassed.

Aristotle, Metaphysics Book 2
*The Complete Works of Aristotle*, Princeton University Press, 1984.

# Preface

Robots manipulating and navigating in unmodelled environments need robust geometric cues to recover scene structure. Vision can provide some of the most powerful cues. However, describing and inferring geometric information about arbitrarily curved surfaces from visual cues is a difficult problem in computer vision. Existing methods of recovering the three-dimensional shape of visible surfaces, e.g. *stereo* and *structure from motion*, are inadequate in their treatment of curved surfaces, especially when surface texture is sparse. They also lack robustness in the presence of measurement noise or when their design assumptions are violated. This book addresses these limitations and shortcomings.

Firstly novel computational theories relating visual motion arising from viewer movements to the *differential geometry* of visible surfaces are presented. It is shown how an *active* monocular observer, making deliberate exploratory movements, can recover reliable descriptions of curved surfaces by tracking image curves. The deformation of *apparent contours* (outlines of curved surfaces) under viewer motion is analysed and it is shown how surface curvature can be inferred from the *acceleration* of image features. The image motion of other curves on surfaces is then considered, concentrating on aspects of surface geometry which can be recovered efficiently and robustly and which are insensitive to the exact details of viewer motion. Examples include the recovery of the sign of *normal curvature* from the image motion of inflections and the recovery of surface orientation and *time to contact* from the *differential invariants* of the image velocity field computed at image curves.

These theories have been implemented and tested using a real-time tracking system based on deformable contours (B-spline snakes). Examples are presented in which the visually derived geometry of piecewise smooth surfaces is used in a variety of tasks including the geometric modelling of objects, obstacle avoidance and navigation and object manipulation.

## Acknowledgements

The work described in this book was carried out at the Department of Engineering Science of the University of Oxford under the supervision of Andrew Blake. I am extremely grateful to him for his astute and incisive guidance and the catalyst for many of the ideas described here. Co-authored extracts from Chapter 2, 3 and 5 have been been published in the International Journal of Computer Vision, International Journal of Robotics Research, Image and Vision Computing, and in the proceedings of the International and European Conferences on Computer Vision. I am also grateful to Andrew Zisserman for his diligent proof reading, technical advice, and enthusiastic encouragement. A co-authored article extracted from part of Chapter 4 appears in the International Journal of Computer Vision.

I have benefited considerably from discussions with members of the Robotics Research Group and members of the international vision research community. These include Olivier Faugeras, Peter Giblin, Kenichi Kanatani, Jan Koenderink, Christopher Longuet-Higgins, Steve Maybank, and Joseph Mundy.

Lastly I am indebted to Professor J.M. Brady, for providing financial support, excellent research facilities, direction, and leadership. This research was funded by the IBM UK Science Centre and the Lady Wolfson Junior Research Fellowship at St Hugh's College, Oxford.

## Dedication

This book is dedicated to my parents, Concetta and Salvatore Cipolla. Their loving support and attention, and their encouragement to stay in higher education (despite the sacrifices that this entailed for them) gave me the strength to persevere.

Cambridge, August 1992                                                          Roberto Cipolla

# Contents

# Chapter 1

# Introduction

## 1.1 Motivation

Robots manipulating and navigating in unmodelled environments need robust geometric cues to recover scene structure. *Vision* – the process of discovering from images what is present in the world and where it is [144] – can provide some of the most powerful cues.

Vision is an extremely complicated sense. Understanding how our visual systems recognise familiar objects in a scene as well as describing qualitatively the position, orientation and three-dimensional (3D) shape of unfamiliar ones, has been the subject of intense curiosity and investigation in subjects as disparate as philosophy, psychology, psychophysics, physiology and artificial intelligence (AI) for many years. The AI approach is exemplified by computational theories of vision [144]. These analyse vision as a complex information processing task and use the precise language and methods of computation to describe, debate and test models of visual processing. Their aim is to elucidate the information present in visual sensory data and how it should be processed to recover reliable three-dimensional descriptions of visible surfaces.

### 1.1.1 Depth cues from stereo and structure from motion

Although visual images contain cues to surface shape and depth, e.g. perspective cues such as vanishing points and texture gradients [86], their interpretation is inherently ambiguous. This is attested by the fact that the human visual system is deceived by "trompe d'oeuil" used by artists and visual illusions, e.g. the Ames room [110, 89], when shown a single image or viewing a scene from a single viewpoint. The ambiguity in interpretation arises because information is lost in the projection from the three-dimensional world to two-dimensional images.

Multiple images from different viewpoints can resolve these ambiguities. Visible surfaces which yield almost no depth perception cues when viewed from a single viewpoint, or when stationary, yield vivid 3D impressions when movement (either of the viewer or object) is introduced. These effects are known as *stereop-sis* (viewing the scene from different viewpoints simultaneously as in binocular vision [146]) and *kineopsis* ( the "kinetic depth" effect due to relative motion between the viewer and the scene [86, 206]). In computer vision the respective paradigms are *stereo vision* [14] and *structure from motion* [201].

In stereo vision the processing involved can be decomposed into two parts.

1. The extraction of disparities (difference in image positions). This involves matching image features that correspond to the projection of the same scene point. This is referred to as the *correspondence problem*. It concerns which features should be matched and the constraints that can be used to help match them [147, 10, 152, 171, 8].

2. The interpretation of disparities as 3D depths of the scene point. This requires knowledge of the camera/eye geometry and the relative positions and orientations of the viewpoints (*epipolar geometry* [10]). This is essentially triangulation of two visual rays (determined by image measurements and camera orientations) and a known baseline (defined by the relative positions of the two viewpoints). Their intersection in space determines the position of the scene point.

Structure from motion can be considered in a similar way to stereo but with the different viewpoints resulting from (unknown) relative motion of the viewer and the scene. The emphasis of structure from motion approach has been to determine the number of (image) points and the number of views needed to recover the spatial configuration of the scene points and the motion compatible with the views [201, 135]. The processing involved can be decomposed into three parts.

1. Tracking features (usually 2D image structures such as points or "corners").

2. Interpreting their image motion as arising from a *rigid* motion in 3D. This can be used to estimate the exact details (translation and rotation) of the relative motion.

3. Image velocities and viewer motion can then be interpreted in the same way as stereo disparities and epipolar geometry (see above). These are used to recover the scene structure which is expressed explicitly as quantitative depths (up to a speed–scale ambiguity).

The computational nature of these problems has been the focus of a significant amount of research during the past two decades. Many aspects are well

(a)

(b)

Figure 1.1: Stereo image pair with polyhedral model.

*The Sheffield Tina stereo algorithm [171] uses Canny edge detection [48] and accurate camera calibration [195] to extract and match 2D edges in the left (a) and right (b) images of a stereo pair. The reconstructed 3D line segments are interpreted as the edges of a polyhedral object and used to match the object to a model database [179]. The models are shown superimposed on the original image (a). Courtesy of I. Reid, University of Oxford.*

(a)

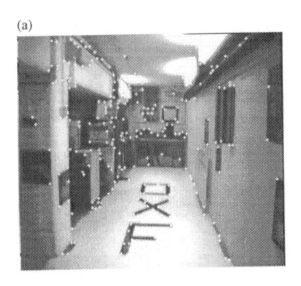

(b)

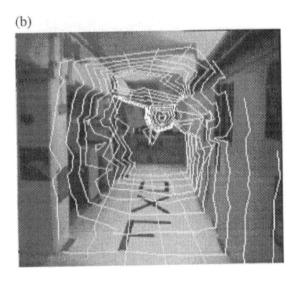

Figure 1.2: Structure from motion.

*(a) Detected image "corners" [97, 208] in the first frame of an image sequence. The motion of the corners is used to estimate the camera's motion (ego-motion) [93]. The integration of image measurements from a large number of viewpoints is used to recover the depths of the scene points [96, 49]. (b) The 3D data is used to compute a contour map based on a piecewise planar approximation to the scene. Courtesy of H. Wang, University of Oxford.*

understood and AI systems already exist which demonstrate basic competences in recovering 3D shape information. The state of the art is highlighted by considering two recently developed and successful systems.

- Sheffield stereo system:
  This system relies on accurate camera calibration and feature (edge) detection to match segments of images edges, permitting recovery 3D line segments [171, 173]. These are either interpreted as edges of polyhedra or grouped into planar surfaces. This data has been used to match to models in a database [179] (figure 1.1).

- Plessey Droid structure from motion system:
  A camera mounted on a vehicle detects and tracks image "corners" over an image sequence. These are used to estimate the camera's motion (egomotion). The integration of image measurements from a large number of viewpoints is used to recover the depths of the scene points. Planar facets are fitted to neighbouring triplets of the 3D data points (from Delaunay triangulation in the image [33]) and their positions and orientations are used to define navigable regions [93, 96, 97, 49, 208] (figure 1.2).

These systems demonstrate that with accurate calibration and feature detection (for stereo) or a wide angle of view and a large range of depths (for structure from motion) stereo and structure from motion are feasible methods of recovering scene structure. In their present form these approaches have serious limitations and shortcomings. These are listed below. Overcoming these limitations and shortcomings – inadequate treatment of curved surfaces and lack of robustness – will be the main themes of this thesis.

## 1.1.2 Shortcomings

### 1. Curved surfaces

Attention to mini-worlds, such as a piecewise planar polyhedral world, has proved to be restrictive [172] but has continued to exist because of the difficulty in interpreting the images of curved surfaces. Theories, representations and methods for the analysis of images of polyhedra have not readily generalised to a piecewise smooth world of curved surfaces.

- **Theory**
  A polyhedral object's line primitives (image edges) are adequate to describe its shape because its 3D surface edges are view-independent. However, in images of curved surface (especially in man-made environments where surface texture may be sparse) the dominant image

line and arc primitives are *apparent contours* (see below). These do
not convey a curved surface's shape in the same way. Their contour generators move and deform over a curved object's surface as
the viewpoint is changed. These can defeat many stereo and structure from motion algorithms since the features (contours) in different
viewpoints are projections of different scene points. This is effectively
introducing non-rigidity.

- **Representation**
  Many existing methods make explicit quantitative depths of visible
  points [90, 7, 96]. Surfaces are then reconstructed from these sparse
  data by interpolation or fitting surface models – the plane being a particularly common and useful example. For arbitrarily curved, smooth
  surfaces, however, no surface model is available that is general enough.

The absence of adequate surface models and the sparsity of surface features make describing and inferring geometric information about 3D curved
objects from visual cues a challenging problem in computer vision. Developing theories and methods to recover reliable descriptions of arbitrarily
curved smooth surfaces is one of the major themes of this thesis.

2. **Robustness**
   The lack of robustness of computer vision systems compared to biological
   systems has led many to question the suitability of existing computational
   theories [194]. Many existing methods are inadequate or incomplete and
   require development to make then robust and capable of recovering from
   error.

   Existing structure from motion algorithms have proved to be of little or
   no practical use when analysing images in which perspective effects are
   small. Their solutions are often ill-conditioned, and fail in the presence of
   small quantities of image measurement noise; when the field of view and
   the variation of depths in the scene is small; or in the presence of small
   degrees of non-rigidity (see Chapter 5 for details). Worst, they often fail
   in particularly graceless fashions [197, 60]. Yet the human visual system
   gains vivid 3D impressions from two views (even orthographic ones) even
   in the presence of non-rigidity [131].

   Part of the problem lies in the way these problems have been formulated.
   Their formulation is such that the interpretation of point image velocities
   or disparities is embroiled in camera calibration and making explicit quantitative depths. Reformulating these problems to make them less sensitive
   to measurement error and epipolar geometry is another major theme of
   this thesis.

## 1.2 Approach

This thesis develops computational theories relating visual motion to the differential geometry of visible surfaces. It shows how an *active* monocular observer can make deliberate movements to recover reliable descriptions of visible surface geometry. The observer then acts on this information in a number of visually guided tasks ranging from navigation to object manipulation.

The details of our general approach are listed below. Some of these ideas have recently gained widespread popularity in the vision research community.

### 1.2.1 Visual motion and differential geometry

Attention is restricted to arbitrarily curved, piecewise smooth (at the scale of interest) surfaces. Statistically defined shapes such as textures and crumpled fractal-like surfaces are avoided. Piecewise planar surfaces are considered as a special case. The mathematics of differential surface geometry [67, 122] and 3D shape play a key role in the derivation and exposition of the theories presented. The deformation of visual curves arising from viewer motion is related to surface geometry.

### 1.2.2 Active vision

The inherent practical difficulties of structure from motion algorithms are avoided by allowing the viewer to make deliberate, controlled movements. This has been termed active vision [9, 2]. As a consequence, it is assumed that the viewer has at least some knowledge of his motions, although this may sometimes be expressed qualitatively in terms of uncertainty bounds [106, 186]. Partial knowledge of viewer motion, in particular constraints on the viewer's translation, make the analysis of visual motion considerably easier and can lead to simple, reliable solutions to the structure from motion problem. By controlling the viewpoint, we can achieve non-trivial visual tasks without having to solve completely this problem.

A moving active observer can also more robustly make inferences about the geometry of visible surfaces by integrating the information from different viewpoints, e.g. using camera motion to reduce error by making repeated measurements of the same features [7, 96, 173]. More important, however, is that controlled viewpoint movement can be used to reduce ambiguity in interpretation and sparsity of data by uncovering desired geometric structure. In particular it may be possible to generate new data by moving the camera so that a contour is generated on a surface patch for which geometrical data is required, thus allowing the viewer to fill in the gaps of unknown areas of the surface. The judicious choice and change of viewpoint can generate valuable data.

### 1.2.3   Shape representation

Listed below are favourable properties desired in a shape descriptor.

1. It should be insensitive to changes in viewpoint and illumination, e.g. invariant measures such as the principal curvatures of a surface patch.

2. It should be robust to noise and resistant to surface perturbations, obeying the principle of graceful degradation:

   ...wherever possible, degrading the data will not prevent delivery of at least some of the answer [144].

3. It should be computationally efficient, the latter being specified by the application.

Descriptions of surface shape cover a large spectrum varying from **quantitative** depth maps (which are committed to a single surface whose depths are specified over a dense grid [90]) to a general **qualitative** description (which are incomplete specifications such as classifying the surface locally as either elliptic, hyperbolic or planar [20]). Different visual tasks will demand different shape descriptors within this broad spectrum. The specification is of course determined by the application. A universal 3D or $2\frac{1}{2}$D sketch [144] is as elusive as a universal structure from motion algorithm.

In our approach we abandon the idea of aiming to produce an explicit surface representation such as a depth map from sparse data [144, 90, 192, 31]. The main drawbacks of this approach are that it is computationally difficult and the fine grain of the representation is cumbersome. The formulation is also naive in the following respects. First, there is no unique surface which is consistent with the sparse data delivered by early visual modules. There is no advantage in defining a *best* consistent surface since it is not clear why a visual system would require such an explicit representation. Direct properties of the surfaces such as orientation or curvature are preferred. Second, the main purpose of surface reconstruction should be to make explicit occlusion boundaries and localise discontinuities in depth and orientation. These are usually more important shape properties than credence on the quality of smoothness.

Qualitative or partial shape descriptors include the incomplete specification of properties of a surface in terms of bounds or constraints; spatial order [213], relative depths, orientations and curvatures; and affine 3D shape (Euclidean shape without a metric to specify angles and distances [131]). These descriptions may superficially seem inferior. They are, however, vital, especially when they can be obtained cheaply and reliably whereas a complete specification of the surface may be cumbersome. It will be shown that they can be used successfully in a variety of visual tasks.

Questions of representation of shape and uncertainty should not be treated in isolation. The specification depends on what the representation is for, and what tasks will be performed with it. Shape descriptions must be useful.

### 1.2.4   Task oriented vision

A key part of the approach throughout this thesis is to test the utility, efficiency and reliability of the proposed theories, methods and shape representations in "real" visual tasks, starting from visual inputs and transforming them into representations upon which reasoning and planning programs act. [1] In this way "action" is linked to "perception". In this thesis visual inferences are tested in a number of visual tasks, including navigation and object manipulation.

## 1.3   Themes and contributions

The two main themes of this thesis are interpreting the images of curved surfaces and robustness.

### 1.3.1   Curved surfaces

Visual cues to curved surface shape include outlines (apparent contour [120]), silhouettes, specularities (highlights [128]), shading and self-shadows [122], cast shadows, texture gradients [216] and the projection of curves lying on surfaces [188]. These have often been analysed in single images from single viewpoints. In combination with visual motion resulting from deliberate viewer motions (or similarly considering the deformations between the images in binocular vision) some of these cues become very powerful sources of geometric information. Surfaces will be studied by way of the image (projection) of curves on surfaces and their deformation under viewer motion. There are two dominant sources of curves in images. The first source occurs at the singularity of the mapping between a patch on the surface and its projection [215]. The patch projects to a smooth piece of contour which we call the *apparent contour* or outline. This occurs when viewing a surface along its tangent plane. The apparent contour is the projection of a fictitious space curve on the surface – the *contour generator* – which separates the surface into visible and occluded parts. Shape recovery from these curves will be treated in Chapter 2 and 3. Image curves also can arise when the mapping from surface to image is not singular. The visual

---

[1]This approach is also known as purposive, animate, behavioural or utilitarian vision.

image of curves or patches on the surface due to internal surface markings or illumination effects is simply a deformed map of the surface patch. This type of image curve or patch will be treated in Chapters 4 and 5.

## 1.3.2   Robustness

This thesis also makes a contribution to achieving reliable descriptions and robustness to measurement and ego-motion errors. This is achieved in two ways. The first concerns sensitivity to image measurement errors. A small reduction in sensitivity can be obtained by only considering features in the image that can be reliably detected and extracted. Image curves (edges) and their temporal evolution have such a property. Their main advantage over isolated surface markings is technological. Reliable and accurate edge detectors are now available which localise surface markings to sub-pixel accuracy [48]. The technology for isolated point/corner detection is not at such an advanced stage [164]. Furthermore, snakes [118] are ideally suited to tracking curves through a sequence of images, and thus measuring the curve deformation. Curves have another advantage. Unlike points ("corners") which only samples the surface at isolated points – the surface could have any shape in between the points – a surface curve conveys information, at a particular scale, throughout its path.

The second aspect of robustness is achieved by overcoming sensitivity to the exact details of viewer motion and epipolar geometry. It will be seen later that point image velocities consist of two components. The first is due to viewer translation and it is this component that encodes scene structure. The other component is due to the rotational part of the observer's motion. These rotations contribute no information about the structure of the scene. This is obvious, since rotations about the optical centres leave the rays, and hence the triangulation, unchanged. The interpretation of point image velocities or disparities as quantitative depths, however, is complicated by these rotational terms. In particular small errors in rotation (assumed known from calibration or estimated from structure from motion) have large effects on the recovered depths.

Instead of looking at point image velocities and disparities (which are embroiled in epipolar geometry and making quantitative depths explicit), part of the solution, it is claimed here, is to look at local, *relative* image motion. In particular this thesis shows that relative image velocities and velocity/disparity gradients are valuable cues to surface shape, having the advantage that they are insensitive to the exact details of the viewer's motion. These cues include:

1. Motion parallax – the relative image motion (both velocities and accelerations) of nearby points (which will be considered in Chapters 2 and 3).

2. The deformation of curves (effectively the relative motion of three nearby points) (considered in Chapter 4).

3. The local distortion of apparent image shapes (represented as an affine transformation) (considered in Chapter 5).

Undesirable global additive errors resulting from uncertainty in viewer motion and the contribution of viewer rotational motion can be cancelled out. We will also see that it is extremely useful to base our inferences of surface shape directly on properties which can be measured in the image. Going through the computationally expensive process of making explicit image velocity fields or attempting to invert the imaging process to produce 3D depths will often lead to ill-conditioned solutions even with regularisation [169].

## 1.4 Outline of book

**Chapter 2** develops new theories relating the visual motion of apparent contours to the geometry of the visible surface. First, existing theories are generalised [85] to show that spatio-temporal image derivatives (up to second order) completely specify the visible surface in the vicinity of the apparent contour. This is shown to be sensitive to the exact details of viewer motion. The relative motion of image curves is shown to provide robust estimates of surface curvature.

**Chapter 3** presents the implementation of these theories and describes results with a camera mounted on a moving robot arm. A computationally efficient method of extracting and tracking image contours based on B-spline snakes is presented. Error and sensitivity analysis substantiate the claims that parallax methods are orders of magnitude less sensitive to the details of the viewer's motion than absolute image measurements. The techniques are used to detect apparent contours and discriminate them from other fixed image features. They are also used to recover the 3D shape of surfaces in the vicinity of their apparent contours. We describe the real-time implementations of these algorithms for use in tasks involving the active exploration of visible surface geometry. The visually derived shape information is successfully used in modelling, navigation and the manipulation of piecewise smooth curved objects.

**Chapter 4** describes the constraints placed on surface differential geometry by observing a surface curve from a sequence of positions. The emphasis is on aspects of surface shape which can be recovered efficiently and robustly and without the requirement of the exact knowledge of viewer motion or accurate image measurements. Visibility of the curve is shown to constrain surface orientation. Further, tracking image curve inflections determines the sign of the normal curvature (in the direction of the surface curve's tangent vector). Examples using this information in real image sequences are included.

**Chapter 5** presents a novel method to measure the *differential invariants* of the image velocity field robustly by computing average values from the integral of normal image velocities around closed contours. This avoids having to recover a dense image velocity field and taking partial derivatives. Moreover integration provides some immunity to image measurement noise. It is shown how an *active* observer making small, deliberate (although imprecise) motions can recover precise estimates of the divergence and deformation of the image velocity field and can use these estimates to determine the object surface orientation and time to contact. The results of real-time experiments in which this visually derived information is used to guide a robot manipulator in obstacle collision avoidance, object manipulation and navigation are presented. This is achieved without camera calibration or a complete specification of the epipolar geometry.

A survey of the literature (including background information for this chapter) highlighting the shortcomings of many existing approaches, is included in Appendix A under bibliographical notes. Each chapter will review relevant references.

# Chapter 2

# Surface Shape from the Deformation of Apparent Contours

## 2.1 Introduction

For a smooth arbitrarily curved surface – especially in man–made environments where surface texture may be sparse – the dominant image feature is the *apparent contour* or silhouette (figure 2.1). The apparent contour is the projection of the locus of points on the object – the *contour generator* or *extremal boundary* – which separates the visible from the occluded parts of a smooth opaque, curved surface.

The apparent contour and its deformation under viewer motion are potentially rich sources of geometric information for navigation, object manipulation, motion-planning and object recognition. Barrow and Tenenbaum [17] pointed out that surface orientation along the apparent contour can be computed directly from image data. Koenderink [120] related the curvature of an apparent contour to the intrinsic curvature of the surface (Gaussian curvature); the sign of Gaussian curvature is equal to the sign of the curvature of the image contour. Convexities, concavities and inflections of an apparent contour indicate, respectively, convex, hyperbolic and parabolic surface points. Giblin and Weiss [85] have extended this by adding viewer motions to obtain quantitative estimates of surface curvature. A surface (excluding concavities in opaque objects) can be reconstructed from the envelope of all its tangent planes, which in turn are computed directly from the family of apparent contours/silhouettes of the surface, obtained under motion of the viewer. By assuming that the viewer follows a *great circle* of viewer directions around the object they restricted the problem of analysing the envelope of tangent planes to the less general one of computing the envelope of a family of lines in a plane. Their algorithm was tested on noise-free, synthetic data (on the assumption that extremal boundaries had been distinguished from other image contours) demonstrating the reconstruction of a planar curve under orthographic projection.

In this chapter this will be extended to the general case of arbitrary non-planar, curvilinear viewer motion under perspective projection. The geometry

(a)

(b)

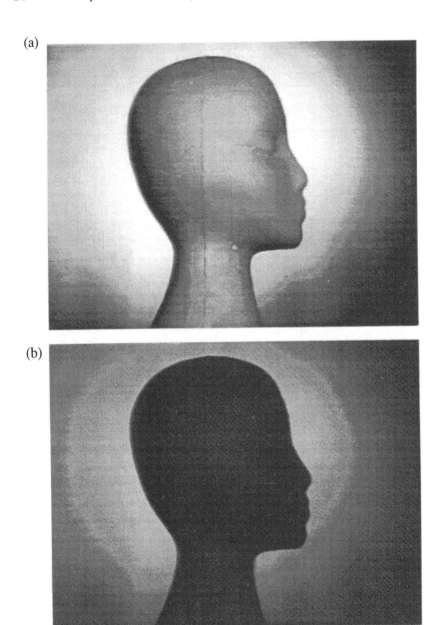

Figure 2.1: A smooth curved surface and its silhouette.

*A single image of a smooth curved surface can provide 3D shape information from shading, surface markings and texture cues (a). However, especially in artificial environments where surface texture may be sparse, the dominant image feature is the outline or apparent contour, shown here as a silhouette (b). The apparent contour or silhouette is an extremely rich source of geometric information. The special relationship between the ray and the local differential surface geometry allow the recovery of the surface orientation and the sign of Gaussian curvature from a single view.*

of apparent contours and their deformation under viewer-motion are related to
the differential geometry of the observed object's surface. In particular it is
shown how to recover the position, orientation and 3D shape of visible surfaces
in the vicinity of their contour generators from the deformation of apparent
contours and *known* viewer motion. The theory for small, local viewer motions
is developed to detect extremal boundaries and distinguish them from *occluding edges* (discontinuities in depth or orientation), surface markings or shadow
boundaries.

A consequence of the theory concerns the robustness of relative measurements of surface curvature based on the relative image motion of nearby points
in the image – *parallax* based measurements. Intuitively it is relatively difficult
to judge, moving around a smooth, featureless object, whether its silhouette is
extremal or not — that is, whether curvature along the contour is bounded or
not. This judgement is much easier to make for objects which have at least a
few surface features. Under small viewer motions, features are "sucked" over the
extremal boundary, at a rate which depends on surface curvature. Our theoretical findings exactly reflect the intuition that the "sucking" effect is a reliable
indicator of relative curvature, regardless of the exact details of the viewer's motion. Relative measurements of curvature across two adjacent points are shown
to be entirely immune to uncertainties in the viewer's rotational velocity.

## 2.2   Theoretical framework

In this section the theoretical framework for the subsequent analysis of apparent
contours and their deformation under viewer motion is presented. We begin
with the properties of apparent contours and their contour generators and then
relate these first to the descriptions of local 3D shape developed from the differential geometry of surfaces and then to the analysis of visual motion of apparent
contours.

### 2.2.1   The apparent contour and its contour generator

Consider a smooth object. For each vantage point all the rays through the vantage point that are tangent to the surface can be constructed. They touch the
object along a smooth curve on its surface which we call the *contour generator* [143] or alternatively the *extremal boundary* [16], the *rim* [120], the *fold* [21]
or the *critical set* of the visual mapping [46, 85] (figure 2.2).

For generic situations (situations which do not change qualitatively under
arbitrarily small excursions of the vantage point) the contour generator is part
of a smooth space curve (not a planar curve) whose direction is not in general
perpendicular to the ray direction. The contour generator is dependent on the

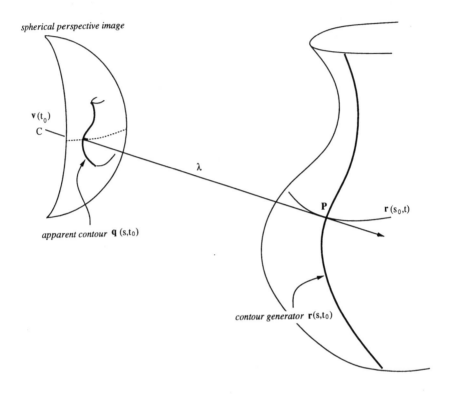

Figure 2.2: Surface and viewing geometry.

*P lies on a smooth surface which is parameterised locally by $\mathbf{r}(s,t)$. For a given vantage point, $\mathbf{v}(t_0)$, the family of rays emanating from the viewer's optical centre (C) that touch the surface defines an s-parameter curve $\mathbf{r}(s,t_0)$ – the contour generator from vantage point $t_0$. The spherical perspective projection of this contour generator – the apparent contour, $\mathbf{q}(s,t_0)$ – determines the direction of rays which graze the surface. The distance along each ray, CP, is $\lambda$.*

local surface geometry and on the vantage point in a simple way which will be elucidated below. Moreover each vantage point will, in general, generate a different contour generator. Movement of the viewer causes the contour generator to "slip" over the visible surface.

The image of the contour generator – here called the apparent contour but elsewhere also known as the occluding contour, profile, outline, silhouette or limb – will usually be smooth (figure 2.2). It may however not be continuous everywhere. As a consequence of the contour generator being a space curve, there may exist a finite number of rays that are tangent not only to the surface but also to the contour generator. At these points the apparent contour of a transparent object will cusp. For opaque surfaces, however, only one branch of the cusp is visible and the contour ends abruptly (see later, figure 2.5) [129, 120].

## 2.2.2   Surface geometry

In the following, descriptions of local 3D shape are developed directly from the differential geometry of surfaces [67, 76, 122].

Consider a point $P$ on the contour generator of a smooth, curved surface in $R^3$ and parameterised locally by a vector valued function $\mathbf{r}(s, t)$. The parametric representation can be considered as covering the surface with two families of curves [134]: $\mathbf{r}(s, t_0)$ and $\mathbf{r}(s_0, t)$ where $s_0$ or $t_0$ are fixed for a given curve in the family. For the analysis of apparent contours and their deformation with viewer motion it is necessary to choose the one-parameter family of views to be indexed by a time parameter $t$, which will also parameterise viewer position for a moving observer. The $s$ and $t$ parameters are defined so that the $s$-parameter curve, $\mathbf{r}(s, t_0)$, is a contour generator from a particular view $t_0$ (figure 2.2). A $t$-parameter curve $\mathbf{r}(s_0, t)$ can be thought of as the 3D locus of points grazed by a light-ray from the viewer, under viewer motion. Such a locus is not uniquely defined. Given a starting point $s = s_0$, $t = t_0$, the correspondence, as the viewer moves, between "successive" (in an infinitesimal sense) contour generators is not unique. Hence there is considerable freedom to choose a spatio-temporal parameterisation of the surface, $\mathbf{r}(s, t)$.

The local surface geometry at $P$ is determined by the tangent plane (surface normal) and a description of how the tangent plane turns as we move in arbitrary directions over the surface (figure 2.3). This can be specified in terms of the basis $\{\mathbf{r}_s, \mathbf{r}_t\}$ for the tangent plane (where for convenience $\mathbf{r}_s$ and $\mathbf{r}_t$ denote $\partial \mathbf{r}/\partial s$ and $\partial \mathbf{r}/\partial t$ – the tangents to the $s$ and $t$-parameter curves respectively) [1]; the surface

---

[1] Subscripts denote differentiation with respect to the subscript parameter. Superscripts will be used as labels.

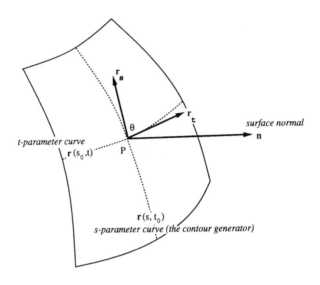

Figure 2.3: The tangent plane.

*Local surface geometry can be specified in terms of the basis $\{\mathbf{r}_s, \mathbf{r}_t\}$ for the tangent plane (where $\mathbf{r}_s$ and $\mathbf{r}_t$ denote the tangents to the s and t-parameter curves respectively and are not in general orthogonal) and the surface normal $\mathbf{n}$ (a unit vector). In differential surface geometry the derivative of these quantities with respect to movement over the surface is used to describe surface shape.*

normal, (a unit vector, $\mathbf{n}$) defined so that

$$\mathbf{r}_s.\mathbf{n} = 0 \tag{2.1}$$

$$\mathbf{r}_t.\mathbf{n} = 0 \tag{2.2}$$

and the derivatives of these quantities with respect to movement over the surface. These are conveniently packaged in the first and second fundamental forms as follows. For a tangent plane vector at $P$, $\mathbf{w}$, the first fundamental form, $I(\mathbf{w}, \mathbf{w})$, is used to express the length of any infinitesimal element in the tangent plane ([67], p.92 ):

$$I(\mathbf{w}, \mathbf{w}) = \mathbf{w}.\mathbf{w}. \tag{2.3}$$

It can be represented by a matrix of coefficients, $\mathbf{G}$, with respect to the basis $\{\mathbf{r}_s, \mathbf{r}_t\}$ where

$$\mathbf{G} = \begin{bmatrix} \mathbf{r}_s.\mathbf{r}_s & \mathbf{r}_s.\mathbf{r}_t \\ \mathbf{r}_t.\mathbf{r}_s & \mathbf{r}_t.\mathbf{r}_t \end{bmatrix}. \tag{2.4}$$

The second fundamental form, $II(\mathbf{w}, \mathbf{w})$, quantifies the "bending away" of the surface from the tangent plane. It is defined by ([67], p.141):

$$II(\mathbf{w}, \mathbf{w}) = -\mathbf{w}.\mathbf{L}(\mathbf{w}) \tag{2.5}$$

where $\mathbf{L}(\mathbf{w})$ is the derivative of the surface orientation, $\mathbf{n}$, in the direction $\mathbf{w}$. $\mathbf{L}$ is in fact a linear transformation on the tangent plane. It is also called the Shape operator [166] or the Weingarten Map [156]. In particular for the basis vectors $\{\mathbf{r}_s, \mathbf{r}_t\}$:

$$\mathbf{L}(\mathbf{r}_s) = \mathbf{n}_s \tag{2.6}$$

$$\mathbf{L}(\mathbf{r}_t) = \mathbf{n}_t \tag{2.7}$$

and the coefficients of the second fundamental are given by matrix $\mathbf{D}$ where

$$\mathbf{D} = \begin{bmatrix} \mathbf{r}_{ss}.\mathbf{n} & \mathbf{r}_{st}.\mathbf{n} \\ \mathbf{r}_{ts}.\mathbf{n} & \mathbf{r}_{tt}.\mathbf{n} \end{bmatrix}. \tag{2.8}$$

The geometry of the surface is completely determined locally up to a rigid motion in $R^3$ by these two quadratic forms. It is, however, sometimes more convenient to characterise the surface by *normal* curvatures in specific directions in the tangent plane [2]. The *normal* curvature in the direction $\mathbf{w}$, $\kappa^n$, is defined by [76]:

$$\kappa^n = \frac{II(\mathbf{w}, \mathbf{w})}{I(\mathbf{w}, \mathbf{w})}. \tag{2.9}$$

---

[2] The *normal* curvature is the curvature of the planar section of the surface through the normal and tangent vector.

The maximum and minimum normal curvatures are called the *principal* curvatures. The corresponding directions are called the *principal* directions [3].

It will now be shown how to make these quadratic forms explicit from image measurable quantities. This requires relating the differential geometry of the surface to the analysis of visual motion.

## 2.2.3   Imaging model

A monocular observer can determine the orientation of any ray projected on to its imaging surface. The observer cannot however, determine the distance along the ray of the object feature which generated it. A general model for the imaging device is therefore to consider it as determining the direction of an incoming ray which we can chose to represent as a unit vector. This is equivalent to considering the imaging device as a spherical pin-hole camera of unit radius (figure 2.2).

The use of spherical projection (rather than planar), which has previously proven to be a powerful tool in structure-from-motion [123] [149], makes it feasible to extend the theory of Giblin and Weiss [85] to allow for perspective. Its simplicity arises from the fact that there are no special points on the image surface, whereas the origin of the perspective plane is special and the consequent loss of symmetry tends to complicate mathematical arguments.

For perspective projection the direction of a ray to a world point, $P$, with position vector $\mathbf{r}(s,t)$, is a unit vector on the image sphere $\mathbf{p}(s,t)$ defined at time $t$ by

$$\mathbf{r}(s,t) = \mathbf{v}(t) + \lambda(s,t)\mathbf{p}(s,t), \qquad (2.10)$$

where $\lambda(s,t)$ is the distance along the ray to the viewed point $P$ and $\mathbf{v}(t)$ is the viewer's position (figure 2.2).

For a given vantage position $t_0$, the apparent contour, $\mathbf{q}(s,t_0)$, determines a continuous family of rays $\mathbf{p}(s,t_0)$ emanating from the camera's optical centre which touch the surface so that

$$\mathbf{p}.\mathbf{n} = 0 \qquad (2.11)$$

where $\mathbf{n}$ is the surface normal. Equation (2.11) defines both the contour generator and the apparent contour.

---

[3] These are in fact the eigenvalues and respective eigenvectors of the matrix $\mathbf{G}^{-1}\mathbf{D}$. The determinant of this matrix (product of the two *principal* curvatures) is called the Gaussian curvature, $K$. It determines qualitatively a surface's shape. A surface patch which is locally hyperbolic (saddle-like) has *principal* curvatures of opposite sign and hence negative Gaussian curvature. Elliptic surface patches (concave or convex) have *principal* curvatures with the same sign and hence positive Gaussian curvature. A locally flat surface patch will have zero Gaussian curvature.

The moving monocular observer at position $\mathbf{v}(t)$ sees a family of apparent contours swept over the image sphere. These determine a two-parameter family of rays in $R^3$, $\mathbf{p}(s,t)$. As before with $\mathbf{r}(s,t)$, the parameterisation is under-determined but that will be fixed later.

### 2.2.4 Viewer and reference co-ordinate systems

Note that $\mathbf{p}$ is the direction of the light ray in the fixed reference/world frame for $R^3$. It is determined by a spherical image position vector $\mathbf{q}$ (the direction of the ray in the camera/viewer co-ordinate system) and the orientation of the camera co-ordinate system relative to the reference frame. For a moving observer the viewer co-ordinate system is continuously moving with respect to the reference frame. The relationship between $\mathbf{p}$ and $\mathbf{q}$ can be conveniently expressed in terms of a rotation operator $R(t)$ [104]:

$$\mathbf{p} = R(t)\mathbf{q}. \tag{2.12}$$

The frames are defined so that instantaneously, at time $t = 0$, they coincide

$$\mathbf{p}(s,0) = \mathbf{q}(s,0) \tag{2.13}$$

and have relative translational and rotational velocities of $\mathbf{U}(t)$ and $\boldsymbol{\Omega}(t)$ respectively:

$$\mathbf{U} = \mathbf{v}_t \tag{2.14}$$

$$(\boldsymbol{\Omega} \wedge \mathbf{q}) = R_t\mathbf{q} \tag{2.15}$$

The relationship between temporal derivatives of measurements made in the camera co-ordinate system and those made in the reference frame is then given by (differentiating (2.12)):

$$\mathbf{p}_t = \mathbf{q}_t + \boldsymbol{\Omega} \wedge \mathbf{q} \tag{2.16}$$

where (as before) the subscripts denote differentiation with respect to time and $\wedge$ denotes a vector product.

## 2.3 Geometric properties of the contour generator and its projection

We now establish why the contour generator is a rich source of information about surface geometry. The physical constraints of *tangency* (all rays at a contour generator are in the surface's tangent plane) and *conjugacy* (the special relationship between the direction of the contour generator and the ray direction) provide powerful constraints on the local geometry of the surface being viewed and allow the recovery of surface orientation and the sign of Gaussian curvature directly from a single image of the contour generator, the apparent contour.

### 2.3.1   Tangency

Both the tangent to the contour generator, $\mathbf{r}_s$ (obtained by differentiating (2.10))

$$\mathbf{r}_s = \lambda_s \mathbf{p} + \lambda \mathbf{p}_s \qquad (2.17)$$

and the ray, $\mathbf{p}$, must (by definition) lie in the tangent plane of the surface. From the tangency conditions

$$\mathbf{r}_s . \mathbf{n} \;=\; 0$$
$$\mathbf{p} . \mathbf{n} \;=\; 0$$

and (2.17), we see that the tangent to the apparent contour also lies in the tangent plane of the surface

$$\mathbf{p}_s . \mathbf{n} = 0. \qquad (2.18)$$

This allows the recovery of the surface orientation $\mathbf{n}$ (defined up to a sign) directly from a single view $\mathbf{p}(s, t_0)$ using the direction of the ray and the tangent to the apparent (image) contour

$$\mathbf{n} = \frac{\mathbf{p} \wedge \mathbf{p}_s}{|\mathbf{p} \wedge \mathbf{p}_s|}. \qquad (2.19)$$

This result is also valid for projection on to the plane. It is a trivial generalisation to perspective projection of the well-known observation of Barrow and Tenenbaum [16, 17].

### 2.3.2   Conjugate direction relationship of ray and contour generator

The *tangency* conditions constrain the contour generator to the tangent plane of the surface. In which direction does the contour generator run? The direction is determined by the second fundamental form and the direction of the ray. In particular the ray direction, $\mathbf{p}$, and the tangent to the contour generator, $\mathbf{r}_s$, are in *conjugate* directions with respect to the second fundamental form [125, 120]. That is, the change in surface normal (orientation of the tangent plane) for an infinitesimal movement in the direction of the contour generator is orthogonal to the direction of the ray. This is intuitively obvious for orthographic projection since the normal will continue to be orthogonal to the line of sight as we move along the contour generator.

This is immediately apparent in the current framework for perspective projection since the second fundamental form has the property that

$$II(\mathbf{p}, \mathbf{r}_s) \;=\; -\mathbf{p} . \mathbf{L}(\mathbf{r}_s)$$
$$= \; -\mathbf{p} . \mathbf{n}_s \qquad (2.20)$$

which, by differentiating (2.11) and substituting (2.18), is zero.

$$\mathbf{p}.\mathbf{n}_s = 0. \tag{2.21}$$

The ray direction, $\mathbf{p}$, and the contour generator are not in general perpendicular but in *conjugate* directions [4].

We will demonstrate the *conjugacy* relationship by way of a few simple examples. Let $\theta$ be the angle between the ray direction $\mathbf{p}$ and the tangent $\mathbf{r}_s$ to the extremal contour. In general $-\pi/2 \leq \theta \leq \pi/2$.

1. $\theta = \pi/2$

   If the ray $\mathbf{p}$ is along a *principal* direction of the surface at P the contour generator will run along the other *principal* direction. *Principal* directions are mutually conjugate. Similarly at an *umbilical* non-planar point, e.g. any point on a sphere, the contour generator will be perpendicular to the ray (figure 2.4a).

2. $-\pi/2 \leq \theta \leq \pi/2$

   At a *parabolic* point of a surface, e.g. any point on a cylinder, the *conjugate* direction of any ray is in the *asymptotic* direction, e.g. parallel to the axis of a cylinder, and the contour generator will then run along this direction and have zero *normal curvature* (figure 2.4b).

3. $\theta = 0$

   The special case $\theta = 0$ occurs when the ray $\mathbf{p}$ lies along an *asymptotic* direction on the surface. The tangent to the contour generator and the ray are parallel – *asymptotic* directions are *self-conjugate*. A *cusp* is generated in the projection of the contour generator, seen as an ending of the apparent contour for an opaque surface[129]  (figure 2.5).

*Conjugacy* is an important relation in differential geometry and vision. As well as determining the direction of a contour generator, it also determines the direction of a self-shadow boundary in relation to its light source [122].

## 2.4   Static properties of apparent contours

It is now well established that static views of extremal boundaries are rich sources of surface geometry [17, 120, 36, 85]. The main results are summarised below followed a description and simple derivation.

---

[4] Since, generically, there is only one direction *conjugate* to any other direction, this property means that a contour generator will not intersect itself.

**(a)**

**(b)**

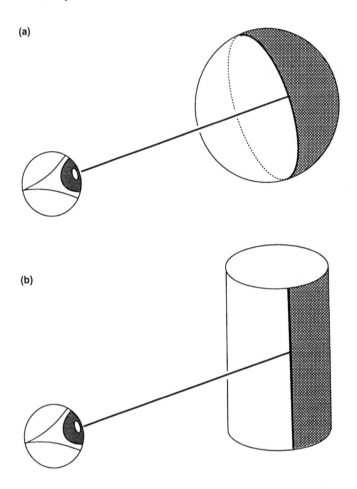

Figure 2.4: In which direction does the contour generator run?

*(a) An example in which the direction of the contour generator is determined by the direction of the ray. For any point on a sphere the contour generator will run in a perpendicular direction to the ray.*

*(b) An example in which the direction of the contour generator is determined by the surface shape. For any point on a cylinder viewed from a generic viewpoint the contour generator will run parallel to the axis and is independent of the direction of the ray.*

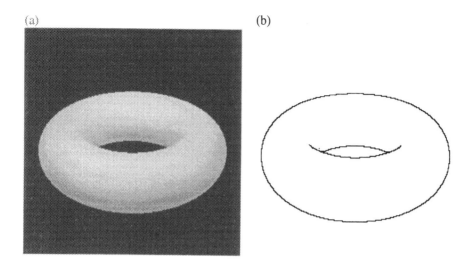

Figure 2.5: Cusps and contour-endings.

*The tangent to the contour generator and the ray are parallel when viewing along an asymptotic direction of the surface. A cusp is generated in the projection of the contour generator, seen as an ending of the apparent contour for an opaque surface. The ending-contour will always be concave. It is however difficult to detect and localise in real images. A synthetic image of a torus (a) and its edges (b) are shown. The edges were detected by a Marr–Hildreth edge finder [78].*

1. The orientation of the tangent plane (surface normal) can be recovered directly from a single view of an apparent contour.

2. The curvature of the apparent contour has the same sign as the normal curvature along the contour generator [5].

3. For opaque surfaces, convexities, inflections and concavities of the apparent contour indicate respectively elliptic, parabolic and hyperbolic surface points.

## 2.4.1  Surface normal

Computation of orientation on a textured surface patch would usually require (known) viewer motion to obtain depth, followed by spatial differentiation. In the case of a contour generator however, the tangency condition (2.11) means that surface orientation $\mathbf{n}(s, t_0)$ can be recovered directly from the apparent contour $\mathbf{p}(s, t_0)$:

$$\mathbf{n}(s, t_0) = \frac{\mathbf{p} \wedge \mathbf{p}_s}{|\mathbf{p} \wedge \mathbf{p}_s|}. \qquad (2.22)$$

The temporal and spatial differentiation that, for the textured patch, would have to be computed with attendant problems of numerical conditioning, is done, for extremal boundaries, by the physical constraint of tangency.

Note that the sign of the orientation can only be determined if it is known on which side of the apparent contour the surface lies. This information may not be reliably available in a single view (figure 2.5b). It is shown below, however, that the "sidedness" of the contour generator can be unambiguously determined from the deformation of the apparent contour under known viewer-motion. In the following we choose the convention that the surface normal is defined to point away from the solid surface. This arbitrarily fixes the direction of increasing s-parameter of the apparent contours so that $\{\mathbf{p}, \mathbf{p}_s, \mathbf{n}\}$ form a right-handed orthogonal frame.

## 2.4.2  Sign of normal curvature along the contour generator

The relationship between the curvature of the apparent contour and the curvature of the contour generator and the viewing geometry is now derived. The curvature of the apparent contour, $\kappa^p$, can be computed as the *geodesic* curvature

---

[5] Note the special case of a cusp when the apparent contour has infinite curvature while the contour generator has zero normal curvature. This case can be considered in the limit.

[6] of the curve, $\mathbf{p}(s, t_0)$, on the image sphere. By definition: [7]

$$\kappa^p = \frac{\mathbf{p}_{ss} . \mathbf{n}}{|\mathbf{p}_s|^2}. \qquad (2.23)$$

It is simply related to the *normal* curvature of the contour generator, $\kappa^s$, by:

$$\kappa^p = \frac{\lambda \kappa^s}{\sin^2 \theta} \qquad (2.24)$$

where (as before) $\theta$ is the angle between the ray and the contour generator,

$$\cos \theta = \mathbf{p} . \frac{\mathbf{r}_s}{|\mathbf{r}_s|} \qquad (2.25)$$

and $\kappa^s$ is the *normal* curvature along the contour generator defined by (2.9):

$$\kappa^s = \frac{\mathbf{r}_{ss} . \mathbf{n}}{\mathbf{r}_s . \mathbf{r}_s}. \qquad (2.26)$$

Since surface depth $\lambda$ must be positive, the sign of $\kappa^s$ must, in fact, be the same as the sign of $\kappa^p$. In the special case of viewing a parabolic point, $\kappa^s = 0$, and an inflection is generated in the apparent contour.

**Derivation 2.1** *The derivation of equation (2.24) follows directly from the equations of perspective projection. Rearranging (2.17) we can derive the mapping between the length of a small element of the contour generator $|\mathbf{r}_s|$ and its spherical perspective projection $|\mathbf{p}_s|$.*

$$|\mathbf{p}_s| = \frac{|\mathbf{r}_s|}{\lambda} \left( 1 - \left( \mathbf{p} . \frac{\mathbf{r}_s}{|\mathbf{r}_s|} \right)^2 \right)^{1/2} \qquad (2.27)$$

$$= \frac{|\mathbf{r}_s|}{\lambda} \sin \theta. \qquad (2.28)$$

*Note that the mapping from contour generator to apparent contour is singular (degenerate) when $\theta$ is zero. The tangent to the contour generator projects to a point in the image. As discussed above this is the situation in which a cusp is generated in the apparent contour and is seen as a contour-ending (figure 2.5).*

---

[6] The *geodesic* curvature of a curve on a sphere is sometimes called the *apparent* curvature [122]. It measures how the curve is curving in the imaging surface. It is equal to the curvature of the perspective projection onto a plane defined by the ray direction.

[7] The curvature, $\kappa$, and the Frenet–Serret normal, $\mathbf{N}$, for a space curve $\gamma(s)$ are given by ([76], p103): $\kappa \mathbf{N} = (\gamma_s \wedge \gamma_{ss}) \wedge \gamma_s / |\gamma_s|^4$. The normal curvature is the magnitude of the component of $\kappa \mathbf{N}$ in the direction of the surface normal (here $\mathbf{p}$ since $\mathbf{p}(s, t_0)$ is a curve on the image sphere); the geodesic curvature is the magnitude of the component in a direction perpendicular to the surface normal and the curve tangent (in this case $\mathbf{p}_s$). For a curve on a sphere this direction is parallel to the curve normal ($\mathbf{n}$ or apparent contours).

*Differentiating (2.17) and collecting the components parallel to the surface normal gives*

$$\mathbf{p}_{ss}.\mathbf{n} = \frac{\mathbf{r}_{ss}.\mathbf{n}}{\lambda}. \tag{2.29}$$

*Substituting (2.27) and (2.29) into the definition of* apparent curvature (2.23) *and normal curvature (2.26) we obtain an alternative form of (2.24):*

$$\kappa^p = \frac{\lambda \kappa^s}{\left[1 - \left(\mathbf{p}.\frac{\mathbf{r}_s}{|\mathbf{r}_s|}\right)^2\right]}. \tag{2.30}$$

A similar result was derived for orthographic projection by Brady et al. [36].

### 2.4.3  Sign of Gaussian curvature

The sign of the Gaussian curvature, $K$, can be inferred from a single view of an extremal boundary by the sign of the curvature of the apparent contour. Koenderink showed that:

> ...from any vantage point and without any restriction on the shape of the rim, a convexity of the contour corresponds to a convex patch of surface, a concavity to a saddle-shaped patch. Inflections of the contour correspond to flexional curves of the surface [120].

In particular he proves Marr wrong:

> In general of course, points of inflection in a contour need have no significance for the surface [144].

by showing that inflections of the contour correspond to parabolic points (where the Gaussian curvature is zero) of the surface.

This follows from a simple relationship between the Gaussian curvature, $K$; the curvature $\kappa^t$ of the normal section at P containing the ray direction; the curvature $\kappa^p$ of the apparent contour (perspective projection) and the depth $\lambda$ [120, 122]:

$$K = \frac{\kappa^p \kappa^t}{\lambda}. \tag{2.31}$$

The sign of $\kappa^t$ is always the same at a contour generator. For $P$ to be visible, the *normal section* must be convex [8] at a contour generator – a concave surface point can never appear on a contour generator of an opaque object. Distance to the contour generator, $\lambda$, is always positive. Hence the sign of $\kappa^p$ determines the sign of Gaussian curvature. Convexities, concavities and inflections of an apparent contour indicate, respectively, convex, hyperbolic and parabolic surface points.

---

[8] If we define the surface normal as being outwards from the solid surface, the *normal* curvature will be negative in any direction for a convex surface patch.

Equation (2.31) is derived in section 2.6.4. An alternative proof of the relationship between the sign of Gaussian curvature and the sign of $\kappa^p$ follows.

**Derivation 2.2** *Consider a tangent vector, $\mathbf{w}$, with components in the basis $\{\mathbf{p}, \mathbf{r}_s\}$ of $(\alpha, \beta)$. Let the normal curvature in the direction $\mathbf{w}$ be $\kappa^n$. From (2.9) its sign is given by:*

$$
\begin{aligned}
\mathrm{sign}(\kappa^n) &= -\mathrm{sign}(\mathbf{w}.\mathbf{L}(\mathbf{w})) \\
&= -(\alpha)^2\mathrm{sign}(\kappa^t) - 2\alpha\beta|\mathbf{r}_s|\mathrm{sign}(\mathbf{p}.\mathbf{L}(\mathbf{r}_s)) - (\beta|\mathbf{r}_s|)^2\mathrm{sign}(\kappa^s) \\
&= -(\alpha)^2\mathrm{sign}(\kappa^t) - (\beta|\mathbf{r}_s|)^2\mathrm{sign}(\kappa^s) \qquad (2.32)
\end{aligned}
$$

*since by the conjugacy relationship (2.21), $\mathbf{p}.\mathbf{L}(\mathbf{r}_s) = 0$. Since the sign of $\kappa^t$ is known at an apparent contour – it must always be convex – the sign of $\kappa^s$ determines the sign of the Gaussian curvature, $K$:*

1. *If $\kappa^s$ is convex all normal sections have the same sign of normal curvature – convex. The surface is locally elliptic and $K > 0$.*

2. *If $\kappa^s$ is concave the sign of normal curvature changes as we change directions in the tangent plane. The surface is locally hyperbolic and $K < 0$.*

3. *If $\kappa^s$ is zero the sign of normal curvature does not change but the normal curvature can become zero. The surface is locally parabolic and $K = 0$.*

*Since the sign of $\kappa^s$ is equal to the sign of $\kappa^p$ (2.24), the curvature of the apparent contour indicates the sign of Gaussian curvature.*

As before when we considered the surface normal, the ability to determine the sign of the Gaussian curvature relies on being able to determine on which side of the apparent contour the surface lies. This information is not readily available from image contour data. It is however available if it is possible to detect a contour-ending since the local surface is then hyperbolic (since the surface is being viewed along an *asymptotic* direction) and the apparent contour must be concave at its end-point [129]. Detecting *cusps* by photometric analysis is a non-trivial exercise (figures 2.5).

## 2.5 The dynamic analysis of apparent contours

### 2.5.1 Spatio-temporal parameterisation

The previous section showed that static views of apparent contours provide useful qualitative constraints on local surface shape. The viewer must however have discriminated apparent contours from the images of other surface curves (such as

surface markings or discontinuities in surface orientation) and have determined
on which side of the image contour the surface lies.

When the viewer executes a *known* motion then surface depth can, of course,
be computed from image velocities [34, 103]. This is correct for static space
curves but it will be shown that it also holds for extremal contour generators
even though they are not fixed in space. Furthermore, if image accelerations
are also computed then full surface curvature (local 3D shape) can be computed
along a contour generator. Giblin and Weiss demonstrated this for orthographic
projection and planar motion [85] (Appendix B). We now generalise these results
to arbitrary non-planar, curvilinear viewer motion and perspective projection.
This requires the choice of a suitable spatio-temporal parameterisation for the
image, $\mathbf{q}(s,t)$, and surface, $\mathbf{r}(s,t)$.

As the viewer moves the family of apparent contours, $\mathbf{q}(s,t)$, is swept out
on the image sphere (figure 2.6). However the spatio-temporal parameterisation
of the family is not unique. The mapping between contour generators, and
hence between apparent contours, at successive instants is under-determined.
This is essentially the "aperture problem" for contours, considered either on the
spherical perspective image $\mathbf{q}(s,t)$, or on the Gauss sphere $\mathbf{n}(s,t)$, or between
space curves on the surface $\mathbf{r}(s,t)$. The choice is arbitrary since the image
contours are projections of different 3D space curves.

## 2.5.2  Epipolar parameterisation

A natural choice of parameterisation (for both the spatio-temporal image and
the surface), is the *epipolar parameterisation* defined by

$$\mathbf{r}_t \wedge \mathbf{p} = 0. \tag{2.33}$$

The tangent to the $t$-parameter curve is chosen to be in the direction of the ray,
$\mathbf{p}$. The physical interpretation is that the grazing/contact point is chosen to
"slip" along the ray. The tangent-plane basis vectors, $\mathbf{r}_s$ and $\mathbf{r}_t$, are therefore in
*conjugate* directions. The advantage of the parameterisation is clear later, when
it leads to a simplified treatment of surface curvature and a unified treatment
of the projection of rigid space curves and extremal boundaries.

A natural *correspondence* between points on successive snapshots of an apparent contour can now be set up. These are the lines of constant $s$ on the image
sphere. Differentiating (2.10) with respect to time and enforcing (2.33) leads to
a "matching" condition

$$\mathbf{p}_t = \frac{(\mathbf{U} \wedge \mathbf{p}) \wedge \mathbf{p}}{\lambda}. \tag{2.34}$$

The corresponding ray in the next viewpoint (in an infinitesimal sense),
$\mathbf{p}(s_0, t + \delta t)$, is chosen so that it lies in the plane defined by $(\mathbf{U} \wedge \mathbf{p})$ – the

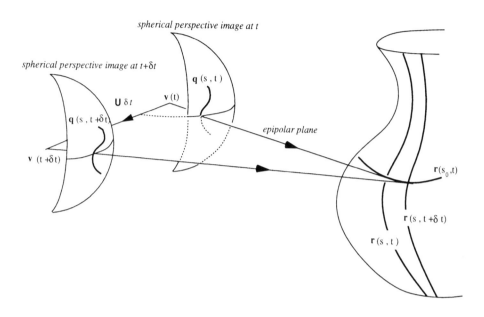

Figure 2.6: Epipolar parameterisation.

*A moving observer at position* $\mathbf{v}(t)$ *sees a family of contour generators* $\mathbf{r}(s, t)$ *indexed by the time parameter* $t$. *Their spherical perspective projections are represented by a two parameter family of apparent contours* $\mathbf{q}(s, t)$. *For the epipolar parameterisation t-parameter curves* $(\mathbf{r}(s_0, t)$ *and* $\mathbf{q}(s_0, t))$ *are defined by choosing the correspondence between successive contours to be in an epipolar plane which is determined by the translational velocity and the direction of the ray.*

*epipolar plane.* The $t$-parameter curve on the surface, $\mathbf{r}(s_0, t)$, will also be in the same plane (figure 2.6).

This is the infinitesimal analogue to epipolar plane matching in stereo [14, 34]. For a general motion, however, the epipolar plane structure rotates continuously as the direction of translation, $\mathbf{U}$, changes and the space curve, $\mathbf{r}(s_0, t)$, generated by the movement of a contact point will be non-planar.

Substituting (2.16) into (2.34), the tangents to the $t$-parameter curves on the spatio-temporal image, $\mathbf{q}(s_0, t)$, are defined by

$$\mathbf{q}_t = \frac{(\mathbf{U} \wedge \mathbf{q}) \wedge \mathbf{q}}{\lambda} - \boldsymbol{\Omega} \wedge \mathbf{q}. \tag{2.35}$$

Note that $\mathbf{q}_t$ is equal to the image velocity of a point on the projection of a static space curve at depth $\lambda$ [149]. This is not surprising since instantaneously image velocities are dependent only on depth and not surface curvature. Points on successive apparent contours are "matched" by searching along epipolar great circles on the image sphere (or epipolar lines for planar image geometry) defined by the viewer motion, $\mathbf{U}$, $\boldsymbol{\Omega}$ and the image position $\mathbf{q}$. This induces a natural correspondence between the contour generators from successive viewpoints on the surface.

The contact point on a contour generator moves/slips along the line of sight $\mathbf{p}$ with a speed, $\mathbf{r}_t$ determined by the distance and surface curvature.

$$\mathbf{r}_t = \left( \frac{\mathbf{p}_t . \mathbf{n}}{\kappa^t} \right) \mathbf{p} \tag{2.36}$$

where $\kappa^t$ is the normal curvature of the space curve, $\mathbf{r}(s_0, t)$:

$$\kappa^t = \frac{\mathbf{r}_{tt} . \mathbf{n}}{\mathbf{r}_t . \mathbf{r}_t}. \tag{2.37}$$

**Derivation 2.3** *Substituting the matching constraint of (2.34) into the time derivative of (2.10) we obtain:*

$$\mathbf{r}_t = (\lambda_t + \mathbf{p} . \mathbf{U}) \mathbf{p}. \tag{2.38}$$

*Differentiating (2.38) with respect to time and substituting this into (2.37) we obtain the relationship between surface curvature and viewer motion.*

$$\kappa^t = \frac{\mathbf{p}_t . \mathbf{n}}{(\lambda_t + \mathbf{p} . \mathbf{U})}. \tag{2.39}$$

*Combining (2.39) and (2.38) gives the required result.*

The numerator of (2.36) is analogous to stereo disparity (as appears below in the denominator of the depth formula (2.40)) and depends only on the distance of the contact point and the "stereo baseline". The denominator is the curvature (normal) of the space curve generated as the viewer moves in time. The speed of the contact point is therefore inversely proportional to the surface curvature. The contour generator "clings" to points with high curvature and speeds up as the curvature is reduced. This property will be used later to distinguish surface markings or creases from extremal boundaries.

## 2.6 Dynamic properties of apparent contours

The choice of a suitable (although arbitrary) spatio-temporal parameterisation permits us to make measurements on the spatio-temporal image, $q(s, t)$, and to recover an exact specification of the visible surface. This includes position, orientation and 3D shape as well as qualitative cues such as to which side of the image contour the occluding surface lies.

### 2.6.1 Recovery of depth from image velocities

Depth $\lambda$ (distance along the ray $p$ — see figure 2.2) can be computed from the deformation ($p_t$) of the apparent contour under known viewer motion (translational velocity $U$)[34]: From (2.34)

$$\lambda = -\frac{U.n}{p_t.n}.$$ (2.40)

This formula is an infinitesimal analogue of triangulation with stereo cameras (figure 2.6). The numerator is analogous to baseline and the denominator to disparity. In the infinitesimal limit stereo will, in principle, correctly determine the depth of the contour generator.

Equation (2.40) can also be re-expressed in terms of spherical image position $q$ and the normal component of image velocity $q_t.n$:

$$\lambda = -\frac{U.n}{q_t.n + (\Omega \wedge q).n}.$$ (2.41)

Clearly absolute depth can only be recovered if rotational velocity $\Omega$ is known.

### 2.6.2 Surface curvature from deformation of the apparent contour

Surface curvature (3D shape) is to be expressed in terms of the first and second fundamental forms, $G$ and $D$ (2.4 and 2.8), which in the epipolar parameterisation and for unit basis vectors can be simplified to:

$$G = \begin{bmatrix} 1 & cos\theta \\ cos\theta & 1 \end{bmatrix}$$ (2.42)

$$\mathbf{D} = \begin{bmatrix} \kappa^s & 0 \\ 0 & \kappa^t \end{bmatrix}, \tag{2.43}$$

where $\kappa^t$ is the normal curvature of the $t$-parameter curve $\mathbf{r}(s_0, t)$ and $\kappa^s$ is the normal curvature of the contour generator $\mathbf{r}(s, t_0)$ at P. Equivalently $\kappa^t$ is the curvature of the normal section at P in the direction of the ray, $\mathbf{p}$.

Note that $\mathbf{D}$ is diagonal. This is a result of choosing, in the epipolar parameterisation, basis directions, $\{\mathbf{r}_s, \mathbf{r}_t\}$ that are *conjugate*. From (2.21) it is easy to show that the off-diagonal components are both equal to zero

$$\mathbf{r}_{ts}.\mathbf{n} = -\mathbf{r}_t.\mathbf{n}_s = -|\mathbf{r}_t|\mathbf{p}.\mathbf{n}_s = 0.$$

How, in summary, can the components of $\mathbf{G}$ and $\mathbf{D}$ be computed from the deformation of apparent contours under viewer motion?

1. **Angle between ray and contour generator, $\theta(s, t_0)$**

   First $\theta(s, t_0)$ can be recovered from the contour generator $\mathbf{r}(s, t_0)$ which is itself obtained from image velocities along the apparent contour via (2.41). This requires the numerical differentiation of depths along the contour generator. From (2.17) and simple trigonometry:

   $$\tan \theta = \frac{\lambda |\mathbf{p}_s|}{\lambda_s} \tag{2.44}$$

2. **Normal curvature along the contour generator, $\kappa^s$**

   Then normal curvature along the contour generator, $\kappa^s$, is computed from the curvature $\kappa^p$ of the apparent contour. Rearranging (2.24):

   $$\kappa^s = \frac{\kappa^p \sin^2 \theta}{\lambda} \tag{2.45}$$

3. **Normal curvature along the line of sight, $\kappa^t$**

   Finally the normal curvature $\kappa^t$, along the line of sight, which can be recovered from image accelerations, as explained below.

   The normal curvature at P in the direction of the ray $\mathbf{p}$, $\kappa^t$ can be computed from the rate of deformation of the apparent contour under viewer motion. From (2.39) and (2.40)

   $$\kappa^t = -\frac{\mathbf{U}.\mathbf{n}}{\lambda(\lambda_t + \mathbf{p}.\mathbf{U})}. \tag{2.46}$$

Since $\kappa^t$ depends on $\lambda_t$, it is clear from (2.41) that $\kappa^t$ is a function of viewer *acceleration* ($\mathbf{U}_t$ and $\mathbf{\Omega}_t$) and the *second* derivative of image position, $\mathbf{q}_{tt}$, that is, image acceleration. By differentiating (2.41) and substituting (2.46) we find that the normal component of image acceleration at an apparent contour is determined by viewer motion (including translational and rotational accelerations) in addition to a dependency on depth and surface curvature:

$$
\mathbf{q}_{tt}.\mathbf{n} = -\frac{(\mathbf{U}.\mathbf{n})^2}{(\lambda)^3}\left[\frac{1}{\kappa^t}\right] - 2\frac{(\mathbf{q}.\mathbf{U})(\mathbf{U}.\mathbf{n})}{(\lambda)^2} - \frac{\mathbf{U}_t.\mathbf{n}}{\lambda} - (\Omega_t \wedge \mathbf{q}).\mathbf{n}
$$
$$
-\frac{2(\mathbf{q}.\mathbf{U})(\Omega \wedge \mathbf{q}).\mathbf{n}}{\lambda} + \frac{(\Omega \wedge \mathbf{U}).\mathbf{n}}{\lambda} + (\Omega.\mathbf{q})(\Omega.\mathbf{n}) \quad (2.47)
$$

The details of equation (2.47) are not important. It merely demonstrates that the recovery of $\kappa^t$ requires knowledge of viewer motion (including translational and rotational accelerations) together with measurement of image accelerations. In section 2.7 it will be shown how to cancel the undesirable dependency on viewer accelerations and rotations.

Note two important points:

- As a result of the *conjugacy* relationship between the contour generator and the ray, surface curvature at a contour generator is completely determined by the *normal* curvatures in these two directions and the angle between them, $\theta$. Compare this to a general surface point which requires the *normal* curvatures in three directions.

- Determining surface curvature usually requires the computation of second-order spatial derivatives of depth, $\lambda$. At extremal boundaries, however, only first order spatial derivatives, $\lambda_s$, and temporal derivatives, $\lambda_t$, need be computed. One derivative is performed, effectively, by the physical system. This is also the case with specularities [26].

## 2.6.3   Sidedness of apparent contour and contour generator

In the static analysis of the apparent contour it was assumed that the "sidedness" of the contour generator – on which side of the image contour the obscuring surface lies – was known. Up to now in the dynamic analysis of apparent contours an arbitrary direction has been chosen for the s-parameter curve (and hence the image tangent $\mathbf{p}_s$) and the surface orientation, $\mathbf{n}$, has been recovered up to an unknown sign from (2.19). The actual sign can now be determined from the deformation of the apparent contour. Equation(2.46) determines both a sign and magnitude for normal curvature along the ray, $\kappa^t$. This must, however, be convex and so its sign given by equation (2.46) allows us to infer the correct orientation of the tangent plane. This is an important qualitative geometric cue. The distinction between free space and solid surface is extremely useful in visual navigation and manipulation.

## 2.6.4   Gaussian and mean curvature

Although the first and second fundamental forms completely characterise the local 3D shape of the surface, it is sometimes more convenient to express the geometry of the surface by its *principal* curvatures and their geometric and arithmetic means: the Gaussian and mean curvature.

The Gaussian curvature, $K$, at a point is given by the product of the two *principal* curvatures [67]. With the epipolar parameterisation, Gaussian curvature can be expressed as a product of two curvatures: the normal curvature $\kappa^t$ and the curvature of the apparent contour, $\kappa^p$ scaled by inverse-depth.

$$K = \frac{\kappa^p \kappa^t}{\lambda} \tag{2.48}$$

This is the well known result of Koenderink [120, 122] extended here to recover the magnitude as well as the sign of Gaussian curvature.

**Derivation 2.4** *In general the Gaussian curvature can be determined from the determinant of $\mathbf{G}^{-1}\mathbf{D}$ or equivalently the ratio of the determinants of the matrices of coefficients of the second and first fundamental forms:*

$$K = \frac{|\mathbf{D}|}{|\mathbf{G}|}. \tag{2.49}$$

*From (2.42) and (2.43) it is trivial to show that Gaussian curvature can be expressed by*

$$K = \frac{\kappa^t \kappa^s}{\sin^2 \theta}. \tag{2.50}$$

*Substituting (2.24) for $\kappa^s$ allows us to derive the result.*

The *mean curvature*, $H$, and the *principal* curvatures $\kappa_1, \kappa_2$ can similarly be expressed by:

$$H = \frac{1}{2} \left[ \frac{\kappa^p}{\lambda} + \kappa^t cosec^2 \theta \right] \tag{2.51}$$

$$\kappa_{1,2} = H \pm \sqrt{H^2 - K}. \tag{2.52}$$

## 2.6.5   Degenerate cases of the epipolar parameterisation

In the previous section we introduced the epipolar parameterisation and showed how to recover the 3D local shape of surfaces from the deformation of apparent contours.

There are two possible cases where degeneracy of the parameterisation arises. These occur when $\{\mathbf{r}_s, \mathbf{r}_t\}$ fails to form a basis for the tangent plane.

1. $\mathbf{r}_t = 0$

   The contour generator does not *slip* over the surface with viewer motion but is fixed. It is therefore not an extremal boundary but a 3D rigid space curve (surface marking or discontinuity in depth or orientation).

   An important advantage of the epipolar parameterisation is its unified treatment of the image motion of curves. The projections of surface markings and creases can be simply be treated as the limiting cases of apparent contours of surfaces with infinite curvature, $\kappa^t$ (from (2.36)). In fact the magnitude of the curvature, $\kappa^t$, can be used to discriminate these image curves from apparent contours. The parameterisation degrades gracefully and hence this condition does not pose any special problems.

   Although the space curve $\mathbf{r}(s, t_0)$ can still be recovered from image velocities via (2.41) the surface orientation is no longer completely defined. The tangency conditions ((2.11) and (2.18)) are no longer valid and the surface normal is only constrained by (2.17) to be perpendicular to the tangent to the space curve, leaving one degree of freedom unknown.

2. $\mathbf{r}_s \wedge \mathbf{r}_t = 0$ and $\mathbf{r}_t \neq 0$

   Not surprisingly, the parameterisation degenerates at the singularity of the surface-to-image mapping where $\mathbf{r}_s$ and $\mathbf{r}_t$ are parallel on the surface. From (2.27) $|\mathbf{p}_s| = 0$ and a cusp is generated in the projection of the contour generator. For generic cases the image contour appears to come to a halt at isolated points.

   Although the epipolar parameterisation and equations (2.19) and (2.40) can no longer be used to recover depth and surface orientation at the isolated cusp point this in general poses no problems. By tracking the contour-ending it is still possible to recover the distance to the surface at a cusp point and the surface orientation [84].

## 2.7 Motion parallax and the robust estimation of surface curvature

It has been shown that although it is feasible to compute surface curvature from the observed deformation of an apparent contour, this requires knowledge of the viewer's translational and rotational velocities and accelerations. Moreover the computation of surface curvature from the deformation of apparent contours is highly sensitive to errors in assumed ego-motion. This may be acceptable for a moving camera mounted on a precision robot arm or when a grid is in view so that accurate visual calibration of the camera position and orientation can be

(a)

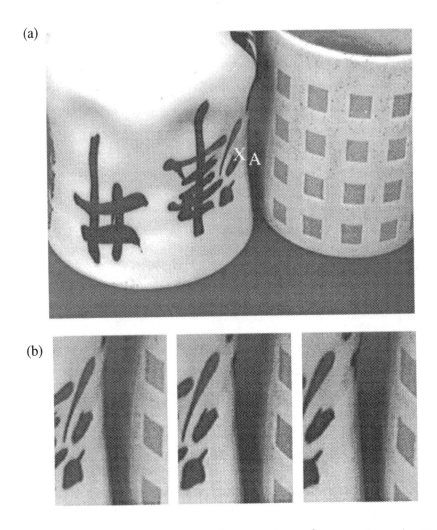

(b)

Figure 2.7: Motion parallax as a robust cue for surface curvature at apparent contours.

*(a) A sample of monocular image sequence showing the image motion of apparent contours with viewer motion.*

*(b) Three images of the sequence (magnified window) showing the relative motion between the apparent contour and a nearby surface markings when the viewer moves from left to right. The relative image accelerations as the features move away from the extremal boundary can be used for the robust estimation of surface curvature.*

(a)

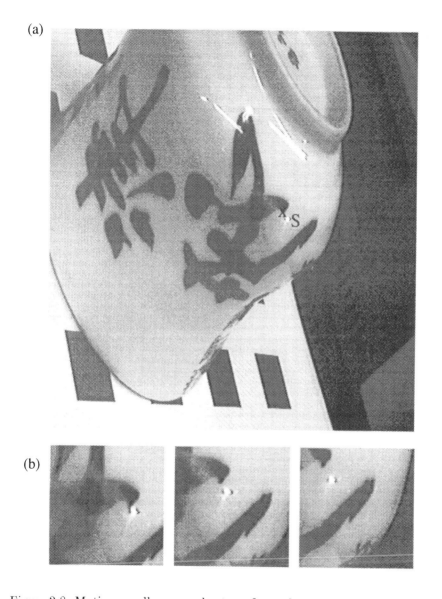

(b)

Figure 2.8: Motion parallax as a robust cue for surface curvature at a specularity.

*(a) Sample from a monocular image sequence showing motion of specularities (highlights) across the curved surface of a Japanese cup with viewer motion (left to right).*
*(b) Three images of the sequence (magnified window) showing the relative motion (parallax) of a point specularity and nearby surface marking. Parallax measurements can be used to determine the surface curvature and normal along the path followed by the specularity as the viewer moves. A more qualitative measure is the sign of the epipolar component of the parallax measurement. With viewer motion the specularity moves in opposite directions for concave and convex surfaces [220]. In the example the specularity moves with the viewer indicating a convex section.*

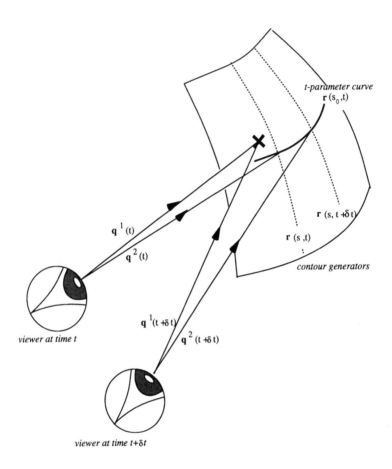

Figure 2.9: Motion parallax.

*Consider the relative displacement between a point on an apparent contour and the image of a nearby surface feature (shown as a cross): $\delta = \mathbf{q}^{(2)} - \mathbf{q}^{(1)}$. The rate of change of relative image position – parallax, $\delta_t$ – has been shown to be a robust indicator of relative depth. In this section we show that its temporal derivative – the rate of parallax $\delta_{tt}$ – is a robust geometric cue for the recovery of surface curvature at extremal boundaries.*

performed [195]. In such cases it is feasible to determine motion to the required accuracy of around 1 part in 1000 (see Chapter 3). However, when only crude estimates of motion are available another strategy is called for. In such a case, it is sometimes possible to use the crude estimate to bootstrap a more precise visual ego-motion computation [93]. However this requires an adequate number of identifiable corner features, which may not be available in an unstructured environment. Moreover, if the estimate is too crude the ego-motion computation may fail; it is notoriously ill-conditioned [197].

The alternative approach is to seek geometric cues that are much less sensitive to error in the motion estimate. In this section it is shown that estimates of surface curvature which are based on the relative image motion of nearby points in the image – *parallax* based measurements – have just this property. Such estimates are stable to perturbations of assumed ego-motion. Intuitively it is relatively difficult to judge, moving around a smooth, featureless object, whether its silhouette is extremal or not — whether curvature along the contour is bounded or not. This judgement is much easier to make for objects which have at least a few surface features. Under small viewer motions, features are "sucked" over the extremal boundary, at a rate which depends on surface curvature (figure 2.7).

The theoretical findings of this section exactly reflect the intuition that the "sucking" effect is a reliable indicator of relative curvature, regardless of the exact details of the viewer's motion. It is shown that relative measurements of curvature across two adjacent points are entirely immune to uncertainties in the viewer's rotational velocity. This is somewhat related to earlier results showing that relative measurements of this kind are important for depth measurement from image velocities[123, 138, 121], or stereoscopic disparities [213] and for curvature measurements from stereoscopically viewed highlights [26] (figure 2.8).

Furthermore, it will be shown that, unlike the interpretation of single-point measurements, differences of measurements at two points are insensitive to errors in rotation and in translational acceleration. Only dependence on translational velocity remains. Typically, the two features might be one point on an extremal boundary and one fixed surface point. The surface point has infinite curvature and therefore acts simply as a stable reference point for the measurement of curvature at the extremal boundary. The reason for the insensitivity of relative curvature measurements is that global additive errors in motion measurement are cancelled out.

## 2.7.1 Motion parallax

Consider two visual features whose projections on the image sphere (figure 2.9) are $\mathbf{q}(s_i, t), i = 1, 2$ (which we will abbreviate to $\mathbf{q}^{(i)}$, $i = 1, 2$) and which have image velocities given by (2.35):

$$q_t^{(i)} = [(U \wedge q^{(i)}) \wedge q^{(i)}] \left[ \frac{1}{\lambda^{(i)}} \right] - \Omega \wedge q^{(i)}. \tag{2.53}$$

Clearly depth can only be recovered accurately if rotational velocity $\Omega$ is known. The dependence on rotational velocity is removed if, instead of using raw image motion $q_t$, the difference of the image motions of a pair of points, $q^{(1)}, q^{(2)}$, is used. This is called *parallax* [99].

The relative image position $\delta$ of the two points is

$$\delta(t) = q^{(2)} - q^{(1)}. \tag{2.54}$$

Parallax is the temporal derivative of $\delta$, $\delta_t$. If instantaneously the two points project to the same point on the image sphere, so that

$$q^{(1)}(0) = q^{(2)}(0) = q,$$

then, from (2.53), the parallax $\delta_t$ depends only on their relative inverse-depths and on viewer velocity. It is independent of (and hence insensitive to errors in) angular rotation $\Omega$:

$$\delta_t = [(U \wedge q) \wedge q] \left[ \frac{1}{\lambda^{(2)}} - \frac{1}{\lambda^{(1)}} \right]. \tag{2.55}$$

The use of "motion parallax" for robust determination of the direction of translation $U$ and relative depths from image velocities was described by Longuet-Higgins and Prazdny [138] and Rieger and Lawton [182].

## 2.7.2   Rate of parallax

Following from the well-known results about motion parallax, we derive the central result of this section — that the *rate* of parallax is a robust cue for surface curvature. The direct formula (2.47) for normal curvature $\kappa^t$ in terms of image acceleration was sensitive to viewer translational acceleration and rotational velocity and acceleration. If, instead, *differences* of image accelerations are used, the undesirable sensitivity vanishes.

The relationship between image acceleration and normal curvature for points $q^{(1)}, q^{(2)}$ can be expressed as

$$\begin{aligned}
q_{tt}^{(i)} . n &= -\frac{(U.n)^2}{(\lambda^{(i)})^3} \left[ \frac{1}{\kappa^{ti}} \right] - 2 \frac{(U.q^{(i)})(U.n)}{(\lambda^{(i)})^2} - \frac{U_t.n}{\lambda^{(i)}} - (\Omega_t \wedge q^{(i)}).n \\
&\quad - \frac{2(U.q^{(i)})(\Omega \wedge q^{(i)}).n}{\lambda^{(i)}} + \frac{(\Omega \wedge U).n}{\lambda^{(i)}} + (\Omega.q^{(i)})(\Omega.n)
\end{aligned} \tag{2.56}$$

The important point is that the two copies of this equation for the two positions $i = 1, 2$ can be subtracted, cancelling off the undesirable dependency on $\Omega, \Omega_t$ and on $\mathbf{U}_t$.

Think of the two points as being the projection of extremal contour generators, which trace out curves with (normal) curvatures $\kappa^{t1}$ and $\kappa^{t2}$ as the viewer moves. Let us define the relative inverse curvature, $\Delta R$, of the feature pair by

$$\Delta R = \frac{1}{\kappa^{t2}} - \frac{1}{\kappa^{t1}}. \tag{2.57}$$

Note that it is simply the difference of the radii of curvature of the normal sections.

$$R^{ti} = \frac{1}{\kappa^{ti}} \qquad i = 1, 2. \tag{2.58}$$

Consider the two features to be instantaneously spatially coincident, that is, initially, $\mathbf{q}(s_1, 0) = \mathbf{q}(s_2, 0)$. Moreover assume they lie at a common depth $\lambda$, and hence, instantaneously, $\mathbf{q}_t^{(1)} = \mathbf{q}_t^{(2)}$. In practice, of course, the feature pair will only coincide exactly if one of the points is a surface marking which is instantaneously on the extremal boundary (figure 2.9). The effect of a small separation is analysed below. Now, taking the difference of equation (2.56) for $i = 1, 2$ leads to the following relation between the two relative quantities $\delta_{tt}$ and $\Delta R$:

$$\delta_{tt}.\mathbf{n} = -\frac{(\mathbf{U}.\mathbf{n})^2}{\lambda^3} \Delta R. \tag{2.59}$$

From this equation we can obtain relative inverse curvature, $\Delta R$, as a function of depth $\lambda$, viewer velocity $\mathbf{U}$, and the second temporal derivative of $\delta$. *Dependence on viewer motion is now limited to the velocity $\mathbf{U}$.* There is no dependence on viewer acceleration or rotational velocity. Hence the relative measurement should be much more robust. (Computationally higher derivatives are generally far more sensitive to noise.)

In the case that $\mathbf{q}^{(1)}$ is known to be a fixed surface reference point, with $1/\kappa^{t1} = 0$, then $\Delta R = 1/\kappa^{t2}$ so that the relative curvature $\Delta R$ constitutes an estimate, now much more robust, of the normal curvature $\kappa^{t2}$ at the extremal boundary point $\mathbf{q}^{(2)}$. Of course this can now be used in equations (2.31), (2.42) and (2.43) to obtain robust estimates of surface curvature. This is confirmed by the experiments of Chapter 3.

Note that the use of the epipolar parameterisation is not important in the above analysis. It can be shown that the normal component of the relative image acceleration $\delta_{tt}.\mathbf{n}$ between a distinct feature and an apparent contour is independent of viewer motion and can be determined completely from spatio-temporal measurements on the image (Appendix C).

### 2.7.3   Degradation of sensitivity with separation of points

The theory above relating relative curvature to the rate of parallax assumed that the two points $\mathbf{q}^{(1)}$ and $\mathbf{q}^{(2)}$ were instantaneously coincident in the image and at the same depth, $\lambda^{(1)} = \lambda^{(2)}$. In practice, point pairs used as features will not coincide exactly and an error limit on curvature (or, more conveniently, its inverse) must be computed to allow for this. The relative curvature can be computed from the rate of parallax by taking into account an error,

$$R^{t(2)} - R^{t(1)} = \Delta R + R^{error} \tag{2.60}$$

where (as before)

$$\Delta R = -\frac{\lambda^3}{(\mathbf{U}.\mathbf{n})^2}\delta_{tt}.\mathbf{n}. \tag{2.61}$$

The error in inverse normal curvature, $R^{error}$, consists of errors due to the difference in depths of the two features, $\Delta\lambda$; the finite separation in the image, $\Delta\mathbf{q}$ and the differences in tangent planes of the two features $\Delta\mathbf{n}$. The magnitude of these effects can be easily computed from the difference of equation (2.56) for the two points (Appendix D). For nearby points that are on the same surface and for fixation ($\mathbf{U} = \lambda\mathbf{\Omega}\wedge\mathbf{q}$) the dominant error can be conveniently expressed as:

$$|R^{error}| \approx 9\lambda|\delta| + \lambda^3|\delta|\frac{|\Omega_t|}{(\mathbf{U}.\mathbf{n})^2} + \lambda|\Delta\lambda|\frac{\mathbf{U}_t.\mathbf{n}}{(\mathbf{U}.\mathbf{n})^2}. \tag{2.62}$$

Parallax-based measurements of curvature will be accurate and insensitive to errors in viewer motion if the separation between points on nearby contours satisfies

$$|\delta| \ll \frac{\Delta R}{9\lambda}. \tag{2.63}$$

Equation (2.62) can also be used to predict the residual sensitivity to translational and rotational accelerations. The important point to notice is that sensitivity to viewer motion is still reduced. As an example consider the sensitivity of absolute measurements of surface curvature along the ray to error in viewer position. Think of this as adding an unknown translational acceleration, $\mathbf{U}_t$. For absolute measurements (2.56) the effect of this unknown error is amplified by a factor of $\lambda^2/(\mathbf{U}.\mathbf{n})^2$ when estimating surface curvature. From Appendix D and (2.62) we see that for parallax-based measurements the sensitivity is reduced to a factor of $2\Delta\lambda/\lambda$ of the original sensitivity. This sensitivity vanishes, of course, when the features are at the same depth. A similar effect is observed for rotational velocities and accelerations.

The residual error and sensitivity analysis can be used to provide an error interval for computed curvature. This is also useful as a threshold for the test (see below) for labelling extremal boundaries — that is, to tell how close to zero the inverse normal curvature $1/\kappa^t$ must lie to be considered to be on a surface marking or a crease edge rather than an extremal boundary.

### 2.7.4  Qualitative shape

Further robustness can be obtained by considering the *ratio* of *relative curvatures*. More precisely this is the ratio of differences in radii of curvature. Ratios of pairs of parallax based measurements can, in theory, be completely insensitive to viewer motion. This is because the normal component of the relative image acceleration $\delta_{tt}.\mathbf{n}$ can be shown to be independent of the viewer motion and can be determined from spatio-temporal measurements on the image for a distinct point and apparent contour pair (Appendix C). This is surprising because the epipolar parameterisation has a hidden dependence on viewer velocity in the "matching" condition (equation(2.34)). This result is important because it demonstrates the possibility of obtaining robust inferences of surface geometry which are independent of any assumption of viewer motion.

In particular if we consider the *ratio* of relative curvature measurements for two *different* point-pairs at similar depths, terms depending on absolute depth $\lambda$ and velocity $\mathbf{U}$ are cancelled out in equation (2.59). This result corresponds to the following intuitive idea. The rate at which surface features rush towards or away from an extremal boundary is inversely proportional to the (normal) curvature there. The constant of proportionality is some function of viewer motion and depth; it can be eliminated by considering only ratios of curvatures.

## 2.8  Summary

This chapter has:

- Related the geometry of apparent contours to the differential geometry of the visible surface and to the analysis of visual motion.

- Shown how the geometric properties of *tangency* and *conjugacy* allow the recovery of qualitative properties of the surface shape from a single view. These include surface orientation and the sign of Gaussian curvature.

- Shown how a moving monocular observer can recover an exact and complete description of the visible surface in the vicinity of a contour generator from the deformation of apparent contours. This requires the computation of spatio-temporal derivatives (up to second order) of the image and known viewer motion. The *epipolar* parameterisation of the spatio-temporal image and surface was introduced. Its advantages include that it allows all image contours to be analysed in the same framework. Image velocities

allow the recovery of the contour generator while image accelerations allow
the computation of surface curvature. A consequence of this is that the
visual motion of curves can be used to detect extremal boundaries and
distinguish them from rigid contour generators such as surface markings,
shadows or creases.

- Shown how the relative motion of image curves (parallax-based measurements) can be used to provide robust estimates of surface curvature which
  are independent (and hence insensitive to) the exact details of the viewer's
  motion.

# Chapter 3

# Deformation of Apparent Contours – Implementation

## 3.1 Introduction

In the previous chapter a computational theory for the recovery of 3D shape from the deformation of apparent contours was presented. The implementation of this theory and the results of experiments performed with a camera mounted on a moving robot arm (figure 3.1) are now described. In particular this chapter presents:

1. A simple, computationally efficient method for accurately extracting image curves from real images and tracking their temporal evolution. This is an extension of tracking with *snakes* [118] — energy minimising splines guided by "image forces" — which avoids computing the internal energies by representing sections of curves as cubic B-splines. Moreover real-time processing (15 frames per second) is achieved by windowing and avoiding Gaussian smoothing.

2. The implementation of the *epipolar* parameterisation to measure image velocities and accelerations. This requires knowledge of the camera motion. Two approaches are presented. The first considers the continuous case and analyses a dense image sequence. Simple linear camera motions and the analysis of epipolar plane slices of the spatio-temporal images are used to estimate the depth and surface curvatures at a point. In the second method we analyse the case of extended discrete displacements and arbitrary known rotations of the viewer. Experiments show that with adequate viewer motion calibration and careful localisation of image contours it is possible to obtain 3D shape measurements of useful accuracy.

3. The analysis of the effect of errors in the knowledge of assumed viewer motion (camera position and orientation) and in the localisation of image contours on the estimates of depth and curvature. Uncertainty and sensitivity analysis is important for two reasons. First, it is useful to

compute bounds on the estimates of surface curvature. This is critical in discriminating fixed features from extremal boundaries – deciding whether curvature along the ray is bounded or not – since with noisy measurements and poorly calibrated viewer motions, we must test by error analysis the hypothesis that the curvature is unbounded at a fixed feature. Second, sensitivity analysis is used to substantiate the claim that *parallax* methods – using the relative image motion of nearby contours – allow the robust recovery of surface curvature. It is shown that estimates of curvature based on absolute measurements of image position are extremely sensitive to motion calibration, requiring accuracies of the order of 1 part in 1000. Estimates of curvature based on *relative* measurements prove to be or-ders of magnitude less sensitive to errors in robot position and orientation. The sensitivity to image localisation remains, however, but is reduced by integrating measurements from a large number of viewpoints.

4. As an illustration of their power, these motion analysis techniques are used to achieve something which has so far eluded analysis based on photometric measurements alone: namely reliable discrimination between fixed surface features and points on extremal boundaries. On which side of the image contour the obscuring surface lies can also be determined. As well as using these methods to detect and label extremal boundaries it is shown how they can recover strips of surfaces in the vicinity of extremal boundaries.

5. The real-time implementation of these algorithms for use in tasks involv-ing the active exploration of the 3D geometry of visible surfaces. This demonstrates the utility and reliability of the proposed theories and meth-ods. It is shown that the deformation of apparent contours under viewer motion is a rich source of geometric information which is extremely useful for visual navigation, motion planning and object manipulation. In these experiments a CCD camera mounted on the wrist joint of a 5-axis Adept 1 SCARA arm has been used (figure 3.1). Examples exploiting the visually derived shape for navigation around and the manipulation of piecewise smooth curved objects are presented.

## 3.2   Tracking image contours with B-spline snakes

Image contours can be localised and tracked using a variant of the well-known "snake" of Kass, Witkin and Terzopoulos [118]. The snake is a computational construct, a dynamic curve able to track moving, deforming image features. Since many snakes can be active at once, each tracking its feature contour as

Figure 3.1: Active exploration of the 3D geometry of visible surfaces.

*In the experiments described in this chapter a single CCD camera mounted on the wrist joint of a 5-axis Adept 1 SCARA arm and controlled by a Sun 4/260 workstation is used to actively recover the geometry of visible surfaces. This information is used in a variety of tasks including 3D object modelling, navigation and manipulation.*

a background process, they constitute a versatile mechanism for direction and focus of attention.

## 3.2.1   Active contours – snakes

Energy-minimising active contour models (snakes) were proposed by Kass et al. [118] as a top-down mechanism for locating features of interest in images and tracking their image motion, provided the feature does not move too fast. The behaviour of a snake is controlled by internal and external "forces" [1]. The internal forces enforce smoothness and the external forces guide the active contour towards the image feature. In their implementation for image curve localisation and tracking, these forces are derived by differentiating internal and external energies respectively.

- **Internal energy**
  The internal energy (per unit length), $E_{internal}$, at a point on the snake, $\mathbf{x}(s)$:

$$E_{internal} = \frac{\alpha |\mathbf{x}_s|^2 + \beta |\mathbf{x}_{ss}|^2}{2} \tag{3.1}$$

  is composed of first and second-order terms, forcing the active contour to act like a string/membrane (avoiding gaps) or a thin rod/plate (avoiding high curvatures) respectively. These effects are controlled by the relative values of $\alpha$ and $\beta$. The internal energy serves to maintain smoothness of the curve under changing external influences.

- **External energy**
  The external force is computed from the image intensity data $I(\mathbf{x}(s))$, where the position of the snake is represented by $\mathbf{x}(s)$, by differentiating an external energy (per unit length) $E_{external}$:

$$E_{external} = -|\nabla G(\sigma) * I(\mathbf{x}(s))|^2 \tag{3.2}$$

  which is computed after convolution of the image with the derivative of a Gaussian kernel, $\nabla G(\sigma)$, of size (scale) $\sigma$. Gaussian smoothing extends the search range of the snake by smearing out image edge features.

The goal is to find the snake (contour) that minimises the total energy. This is achieved by the numerical solution of the elastic problem using techniques from variational calculus. The main step is the solution of a linear equation involving a banded matrix, typically in several hundred variables [118].

Kass et al. [118] describe experiments with snakes in an interactive human machine environment, with the user supplying an initial estimate of the feature

---

[1] The forces are derived from an arbitrary field. They are not natural forces.

and with the snake accurately localising and tracking it. Amini et al. [3] have discussed the problems with the approach. These include instability and a tendency for points to bunch up on strong portions of an edge. They present an implementation based on dynamic programming instead of variational methods which allows the inclusion of hard constraints (which may not be violated) as well as the original smoothness constraints (which do not have to be satisfied exactly). Their approach uses points on a discrete grid and is numerically stable. Hard constraints also allow the distance between points on the snake to be fixed and hence can avoid bunching. However the main drawback is that the method is slow since it depends on the number of the sample points and the cube of the search space.

## 3.2.2 The B-spline snake

A more economical realisation can be obtained by using far fewer state variables [184]. In [50] cubic B-splines [76] were proposed. These are deformable curves represented by four or more state variables (control points). The curves may be open or closed as required. The flexibility of the curve increases as more control points are added; each additional control point allows either one more inflection in the curve or, when multiple knots are used [18], reduced continuity at one point.

B-spline snakes are ideally suited for representing, detecting and tracking image curves. Their main advantages include:

- Local control — modifying the position of a data-point or control point causes only a small part of the curve to change;

- Continuity control — B-splines are defined with continuity properties at each point;

- Compact representation — the number of variables to be estimated is reduced to the number of control-points [154].

The B-spline is a curve in the image plane (figure 3.2)

$$\mathbf{x}(s) = \sum_i f_i(s) \mathbf{Q}_i \qquad (3.3)$$

where $f_i$ are the spline basis functions and $\mathbf{Q}_i$ are the coefficients or control points. These are positioned so that the curve locates the desired image contour. In the original implementation the "external force" on a point $\mathbf{x}(s_j)$ was chosen to be

$$\mathbf{F}(s_j) = \nabla |\nabla G(\sigma) * I(\mathbf{x}(s_j))| \qquad (3.4)$$

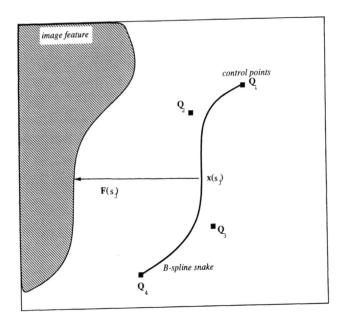

Figure 3.2: The B-spline snake.

*The B-spline snake can be used for image contour localisation and representation. A cubic B-spline can be represented by a minimum of 4 control points, $\mathbf{Q}_i$, and these are positioned so as to locate a nearby contour. The snake moves under the influence of external forces, $\mathbf{F}(s_j)$, which guide it towards the image feature.*

so that, at equilibrium (when image forces vanish), the B-spline, $\mathbf{x}(s)$, stabilised close to a high-contrast contour.

In this section two major simplifications are introduced. The first concerns the control of spatial scale for tracking. The other simplification is that there are no internal forces since the B-spline representation maintains smoothness via hard constraints implicit in the representation.

## External forces and control of scale

The "force" is chosen to depend on the distance between the feature of interest (an edge) and the approximation by the B-spline. For each sample point the "force" on the snake is found by a coarse-to-fine strategy. This is done by inspecting intensity gradients on either side of the snake (either along the normal or along a direction determined by hard constraints, e.g. scan-lines). Control of scale is achieved by inspecting gradients nearer to or further from the snake itself. Each point chooses to move in the direction of the largest intensity gradient (hence towards a contrast edge). If the intensity gradients either side of the contour have opposite signs the scale is halved. This is repeated until the edge has been localised to the nearest pixel.

The gradient is estimated by finite differences. Gaussian smoothing is not used [2]. Image noise is not, as might be thought, problematic in the unblurred image since CCD cameras have relatively low signal-to-noise. Moreover, gradients are sampled at several places along the spline, and those samples combined to compute motions for the spline control points (described below, (3.5) and (3.10)). The combination of those samples itself has an adequate averaging, noise-defeating effect.

## Positioning the control points

External forces are sampled at $N$ points,

$$\mathbf{x}(s_j), \quad j = 1, ..N,$$

along the curve – typically $N > 20$ has been adequate in our experiments. External forces are applied to the curve itself but for iterative adjustment of displacement it is necessary to compute the force transmitted to each control point. This can be done in one of two ways.

In the *first* method this is achieved via the principle of virtual work. At each

---

[2] Small amounts of smoothing can be achieved economically by defocusing the image until the desired degree of blur is achieved. Whilst this worked satisfactorily, it was found that the tracker continued to operate just as well when the lens was sharply focused.

iteration the movement of the control points, $\Delta \mathbf{q}_i$ is given by

$$\Delta \mathbf{Q}_i = \alpha \sum_j f_i(s_j) \mathbf{F}(s_j) \tag{3.5}$$

where $\alpha$, the compliance constant, is chosen so that, in practice, the maximum movement at any iteration always lies within scale of interest. Note that the number of variables to be estimated is reduced to the number of control points.

**Derivation 3.1** *If the control points are displaced by $\Delta \mathbf{Q}_i$, a point on the curve at $\mathbf{x}(s_j)$ is displaced by $\Delta \mathbf{x}(s_j)$ and the work done, $\Delta E$, is approximately:*

$$\Delta E = - \sum_{j=0}^{N} \Delta \mathbf{x}(s_j).\mathbf{F}(s_j). \tag{3.6}$$

*Substituting for the derivative of the equation of the B-spline (3.3)*

$$\begin{aligned}
\Delta E &= - \sum_j (\sum_i f_i(s_j)\Delta \mathbf{Q}_i).\mathbf{F}(s_j) \\
&= - \sum_i \Delta \mathbf{Q}_i.\mathbf{P}_i \tag{3.7}
\end{aligned}$$

*where,*

$$\mathbf{P}_i = \sum_j f_i(s_j)\mathbf{F}(s_j) \tag{3.8}$$

*is the effective force on the $i^{\text{th}}$ control point. The potential energy of the system is minimised by moving the control points parallel to $\mathbf{P}_i$ and so their motion is given by (3.5).*

An alternative method is to position the B-spline so that it minimises the sum of the square of the distances between the discrete data points of the feature and the approximation by the B-spline. Effectively, each snake point is attached to a feature by an elastic membrane so that its potential energy is proportional to the distance squared. This technique has been used to represent image curves [155].

**Derivation 3.2** *If the desired feature position is given by $\mathbf{y}(s_j)$ for a point on the B-spline, $\mathbf{x}(s_j)$, we wish to minimise the potential energy:*

$$\sum_j \left( \mathbf{y}(s_j) - \sum_i f_i(s_j)\mathbf{Q}_i \right)^2. \tag{3.9}$$

*The new positions of the control points, $\mathbf{Q}_i$, are chosen by solving (the least squares solution):*

$$\sum_i \mathbf{Q}_i \sum_j f_i f_k = \sum_j f_k \mathbf{y}(s_j), \tag{3.10}$$

*where $k$ has the same range of values as the control points, $i$.*

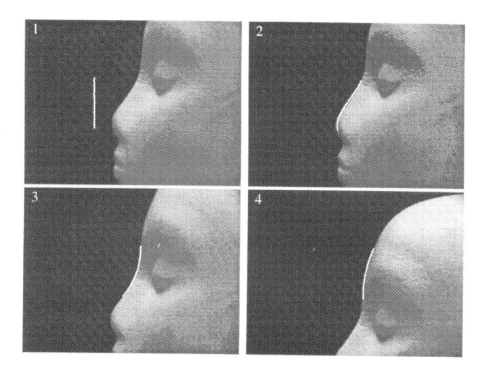

Figure 3.3: Tracking image contours with B-spline snakes.

*A single span B-spline snake "hangs" in the image until it is swept by the motion of the camera into the vicinity of a high contrast edge (top left). The snake then tracks the deforming image contour as the camera is moved vertically upwards by the robot. Four samples of an image sequence are shown in which the robot moves with a speed of 20mm/s. Tracking speeds of 15Hz have been achieved without special purpose hardware by windowing and avoiding Gaussian smoothing.*

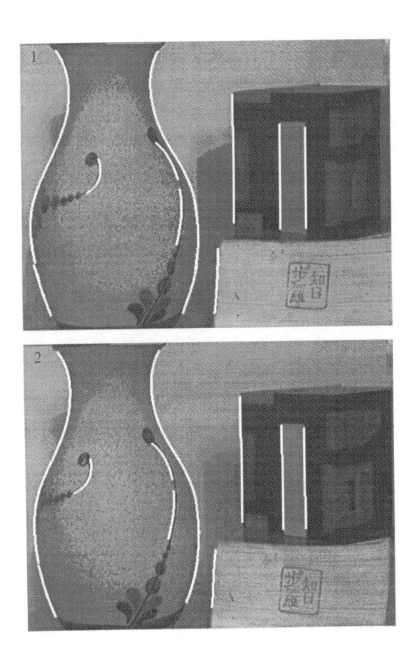

Figure 3.4: Localising and tracking image contours.

*Ten multi-span B-spline snakes were initialised by hand in the first frame near image contours of interest. After localising the contours they were able to track them automatically over the image sequence.*

As the snake approaches the image contour it "locks on" and the scale is reduced to enable accurate contour localisation. Since accurate measurements are required to compute image accelerations, care has been taken over sub-pixel resolution. At earlier stages of tracking, when coarse blurring (large scale) is used, the capture range of the snake is large but localisation is poor – the snake may lag behind the contour. Once the snake has converged on to the contour, standard edge-detection techniques such as smoothing for sub-pixel resolution [48] are used to obtain accurate localisation.

The snakes were initialised by hand in the first frame near images contours of interest after which they track the image contour automatically. Experiments in which the snakes wait in the image until they are swept by the motion of the camera over a feature for which they have an affinity have also been successful. Tracking is maintained provided the contour does not move too quickly. Examples are shown in figures 3.3 and 3.4. The contour tracker can run at 15Hz on a SUN4/260.

Current work at Oxford is developing a real-time tracking system with computer control of focus, zoom and camera motion. Curwen et al. [59] show how interframe constraints can be used to enhance the tracking capability of the B-spline snake by simulating inertia (so that the snake prefers to move in a continuous fashion) and damping (to avoid oscillations). Their work demonstrates that the inclusion of dynamic properties greatly enhances tracking performance. Using a parallel MIMD architecture (based on Transputers) they have achieved panning velocities of $180^o$/second and accelerations as high as $240^o$/second$^2$. Snakes as long as 10 spans can run in real-time on a 9-Transputer system.

In the following section B-spline snakes are used to represent and measure velocities and accelerations at image contours. In later chapters (see Chapters 4 and 5) the real-time B-spline snake will be used for tracking and in the analysis of visual motion.

## 3.3  The epipolar parameterisation

In the *epipolar* parameterisation of the spatio-temporal image and surface, a point on an apparent contour in the first image is "matched" to a point in successive images (in an infinitesimal sense) by searching along the corresponding epipolar lines. This allows us to extract a $t$-parameter curve from the spatio-temporal image. As shown in the previous chapter, depth and surface curvature are then computed from first and second-order temporal derivatives of this $t$-parameter image curve by equations (2.40) and (2.47). This is a non-trivial practical problem since the epipolar structure is continuously changing for arbitrary viewer motions. It requires a dense image sequence and knowledge of the geometrical and optical characteristics of the camera (the *intrinsic* parameters – e.g. image centre, pixel size and focal length [75, 82, 199]) as well as the camera motion.

Estimates of the camera motion are either determined directly from the position and orientation of the gripper and its relationship with the camera centre [198] or are obtained by visual calibration techniques [195]. Extraction of the $t$-parameter curve can be done in a number of ways. We present two simple methods. The first is an extension of epipolar plane image analysis [34, 218] and allows the recovery of depth and surface curvature at a point. The second method analyses the case of extended displacements and arbitrary rotations of the viewer to recover constraints on surface curvature.

## 3.3.1  Epipolar plane image analysis

The *epipolar* parameterisation of the image is greatly simplified for simple motions. In particular if we consider linear viewer motion perpendicular to the optical axis, epipolar lines are simply corresponding raster lines of subsequent images. Figure (3.5) shows the spatio-temporal image formed by taking a sequence of images in rapid succession and stacking these sequentially in time. For linear motions of the viewer, the $t$-parameter image curves are trajectories lying in horizontal slices of the spatio-temporal image. Each horizontal slice corresponds to a different epipolar plane. The trajectories of the image positions of points on an apparent contour (A) and a nearby surface marking (B) are shown as a function of time in the spatio-temporal cross-section of figure 3.5 and plotted in figure 3.6. Note that this is a simple extension of epipolar plane image analysis in which the trajectories of fixed, rigid features appear as straight lines in the spatio-temporal cross-section image with a gradient that is proportional to inverse depth [34]. For apparent contours however, the trajectories are no longer straight. It shown below that the gradient of the trajectory still encodes depth. The curvature determines the curvature of the surface in the epipolar plane.

### Estimation of depth and surface curvature

For motion perpendicular to the optical axis and for an apparent contour which at time $t$ is instantaneously aligned with the optical axis (image position in mm $X(0) = 0$) it is easy to show (from (2.35) and (2.47)) that the image velocity, $X_t(0)$, and acceleration, $X_{tt}(0)$, along the scan-line are given by

$$X_t(0) \quad = \quad -\frac{fU}{\lambda} \tag{3.11}$$

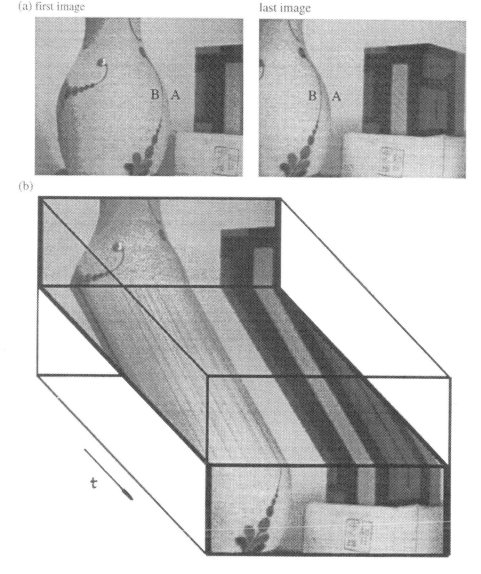

Figure 3.5: 3D spatio-temporal image.

*(a) The first and last image from an image sequence taken from a camera mounted on a robot arm and moving horizontally from left to right without rotation.*

*(b) The 3D spatio-temporal image formed from the image sequence piled up sequentially with time. The top of the first image and the bottom of the last image are shown along with the spatio-temporal cross-section corresponding to the same epipolar plane. For simple viewer motions consisting of camera translations perpendicular to the optical axis the spatio-temporal cross-section image is formed by storing the scan-lines (epipolar lines) for a given epipolar plane sequentially in order of time.*

$$X_{tt}(0) = -\frac{fU^2R}{\lambda^3}, \tag{3.12}$$

where $U$ is the component of viewer translational velocity perpendicular to the optical axis and parallel to the scan line; $f$ is the focal length of the CCD camera lens and $R$ is the radius of curvature of the $t$-parameter curve lying in the epipolar plane. The radius of curvature is related to the curvature of the normal section by Meusnier's formula [67]. The normal curvature along the ray is given by

$$\kappa^t = \frac{1}{R}\cos\phi \tag{3.13}$$

where $\cos\phi$ is the angle between the surface normal and the epipolar plane. In this case, it is simply equal to the angle between the image curve normal and the horizontal scan line.

The estimates of depth and surface curvature follow directly from the first and second temporal derivatives of the $t$-parameter curve, $X(t)$. Due to measurement noise and vibrations of the robot arm, the trajectory may not be smooth and so these derivatives are computed from the coefficients of a parabola fitted locally to the data by least squares estimation. The uncertainty due to random image localisation and ego-motion errors can be derived from analysis of the residual errors.

**Derivation 3.3** *Consider $L$ measurements of image position $X_i$ at times $t_i$. Assume these can be approximated by*

$$X_i = a_0 + b_0 t_i + c_0 t_i^2 + \epsilon_i, \tag{3.14}$$

*where $\epsilon_i$ is a measurement error. Writing this for $L$ points (where $L$ is greater than 3) the coefficients of the parabola can be obtained by solving the following system of linear equations*

$$\begin{bmatrix} 1, t_i, t_i^2 \end{bmatrix} \begin{bmatrix} a_0 \\ b_0 \\ c_0 \end{bmatrix} + [\epsilon_i] = [X_i] \tag{3.15}$$

*or in matrix form*

$$[A]\alpha + \eta = \mathbf{b} \tag{3.16}$$

*where $\alpha$ is the vector of parameters to be estimated (the coefficients of the parabola), $\eta$ is the vector of errors and $\mathbf{b}$ is the vector of measurements (image positions). If $[A^T A]$ is non-singular, the least squares estimate of the parameter vector, $\hat{\alpha}$ is:*

$$\hat{\alpha} = [A^T A]^{-1} A^T \mathbf{b}. \tag{3.17}$$

*The variance–covariance of the parameter vector is given by*

$$Var(\hat{\alpha}) = [A^T A]^{-1} \sigma^2, \tag{3.18}$$

*where $\sigma^2$ is the variance of the measurement error. Since $\sigma^2$ is not known a priori it must be estimated from the residual errors*

$$V^2 = \| A\hat{\alpha} - \mathbf{b} \|^2 . \tag{3.19}$$

*An unbiased estimator for it, $\hat{\sigma}^2$, is given by [178]*

$$\hat{\sigma}^2 = \frac{V^2}{(L-3)}. \tag{3.20}$$

Equations (3.11, 3.12) and the coefficients of the parabola are used to estimate the depth, $\lambda$ and radius of curvature, $R$. The variance–covariance matrix of the parameter vector can be used to compute uncertainty bounds on these estimates.

In practice the viewer will not execute simple translational motions perpendicular to the optical axis but will rotate to fixate on an object of interest. For linear translational viewer motions with known camera rotations the analysis of epipolar plane images is still appropriate if we rectify the detected image curves. Rectification can be performed by a $3 \times 3$ rotation matrix relating measurements in the rotated co-ordinate frame to the standard parallel geometry frame.

For arbitrary curvilinear motions the $t$-parameter curves are no longer constrained to a single cross-section of the spatio-temporal image. Each time instant requires a different epipolar structure and so extracting the $t$-parameter curve from the spatio-temporal image poses a more difficult practical problem.

### Experimental results – curvature from the spatio-temporal image

Figure 3.6 shows the $t$-parameter trajectories for both a feature on an apparent contour (A) and a nearby surface marking (B). The trajectories are both approximately linear with a gradient that determines the distance to the feature. Depth can be estimated to an accuracy of 1 part in 1000 (table 3.1).

The effect due to surface curvature is very difficult to discern. This is easily seen however if we look at the deviation of image position away from the straight line trajectory of a feature at a fixed depth (figure 3.7). Notice that the image position is noisy due to perturbations in the robot position. Typically the robot vibrations have amplitudes between 0.1mm and 0.2mm. From (2.47) we see that these vibrations are amplified by a factor depending on the square of the distance to the feature, and that this results in a large uncertainty in the estimate of surface curvature. Equations (3.11) and (3.12) are used to estimate the depth

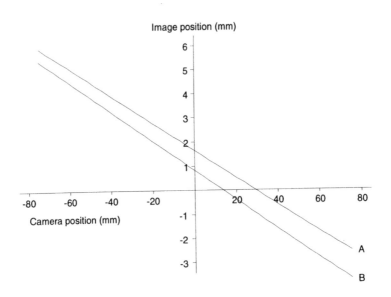

Figure 3.6: Spatio-temporal cross-section image trajectories.

*For linear motion and epipolar parameterisation the t-parameter surface curves lie in the epipolar plane. The t-parameter spatio-temporal image trajectory is also planar. The gradient and curvature of this trajectory encode depth to the contour generator and curvature in the epipolar plane respectively.*

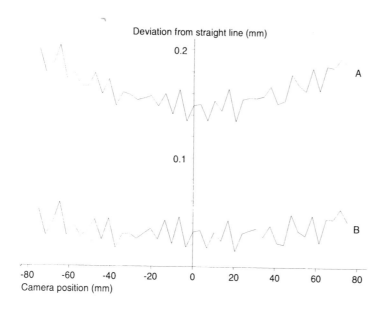

Figure 3.7: Deviation from the straight line trajectory.

*The curvature of the spatio-temporal trajectories is used to estimate the curvature of the epipolar section. The trajectories are not smooth due to vibrations of the robot manipulator (amplitude 0.2mm). Their effect on the estimation of curvature is reduced by a least squares fit to the data. The surface curvatures at A and B are estimated as $51.4 \pm 8.2mm$ and $11.8 \pm 7.3mm$ respectively. B is not on an extremal boundary but is on a fixed curve. This is a degenerate case of the parameterisation and should ideally have zero "radius of curvature", i.e. the spatio-temporal trajectory should be a straight line.*

| | measured depth | measured curvature | actual curvature |
|---|---|---|---|
| extremal boundary, A | 424.3 ± 0.5mm | 51.4 ± 8.2mm | 37 ± 2mm |
| surface marking, B | 393.9 ± 0.4mm | 11.8 ± 7.3mm | 0 |
| parallax measurement (A) | 424.3 ± 0.5mm | 39.6 ± 2mm | 37 ± 2mm |

Table 3.1: Radius of curvature of the *epipolar section* estimated from the spatio-temporal trajectory for a point on an extremal boundary (A) and on a surface marking (B).

and curvature for a point on the extremal boundary of the vase (A) by fitting a parabola to the spatio-temporal trajectory. The method is repeated for a point which is not on an extremal boundary but is on a nearby surface marking (B). This is a degenerate case of the parameterisation. A surface marking can be considered as the limiting case of a point with infinite curvature and hence ideally will have zero "radius of curvature". The estimates of depth and curvature are shown in table 3.1. The veridical values of curvature were measured using calipers. Note that there is a systematic error, not explained by the random errors in the data. This is possibly due to an error in the assumed ego-motion of the robot or focal length.

Figure 3.8 plots the relative image position between A and B against robot position (time). The curvature of this parabola also encodes the surface curvature at A. The parabola is considerably smoother since the effects of robot "wobble" are attenuated when making relative measurements. This is because *the amplification of robot vibrations is reduced* by an order of magnitude. The exact factor depends on the difference of depths between the two features, as predicted by (2.62). In this experiment this corresponds to an order of magnitude. This results in a greatly reduced uncertainty in the estimate of relative inverse curvature. Better still, the estimate of surface curvature based on relative measurements is also more accurate. As predicted by the theory, the nearby point B acts as a stable reference point. Global additive errors in the robot motion effect both the visual motion of both A and B and hence can be cancelled out when the differences of positions are used.

### 3.3.2   Discrete viewpoint analysis

The previous method required a dense (continuous) image sequence. The information available from extended displacements (and arbitrary rotations) of the viewer is now analysed. We will show that from three discrete views it is possible to determine whether or not a contour is extremal. For a surface marking

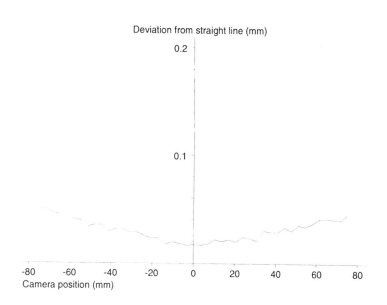

Figure 3.8: Relative image positions.

*The effect of robot vibrations is greatly reduced if relative image positions are used instead.*

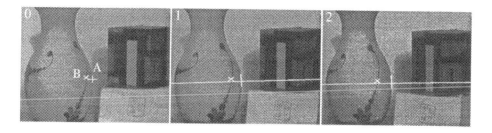

Figure 3.9: Estimating surface curvatures from three discrete views.

*Points are selected on image contours in the first view ($t_0$), indicated by crosses A and B for points on an extremal boundary and surface marking respectively. For epipolar parameterisation of the surface corresponding features lie on epipolar lines in the second and third view ($t_1$ and $t_2$). Measurement of the three rays lying in an epipolar plane can be used to estimate surface curvatures (figure 3.10).*

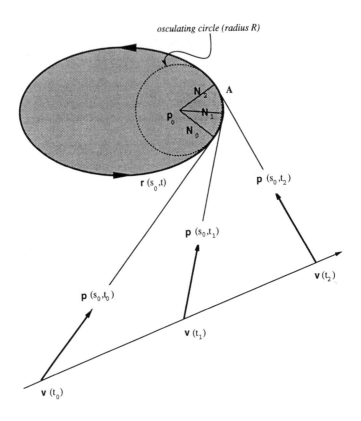

Figure 3.10: The epipolar plane.

*Each view defines a tangent to $\mathbf{r}(s_0, t)$. For linear camera motion and epipolar parameterisation the rays and $\mathbf{r}(s_0, t)$ lie in a plane. If $\mathbf{r}(s_0, t)$ can be approximated locally as a circle, it can be uniquely determined from measurements in three views.*

or crease (discontinuity in surface orientation) the three rays should intersect at point in space for a static scene. For an extremal boundary, however, the contact point slips along a curve, $\mathbf{r}(s_0, t)$ and the three rays will not intersect (figure 3.9 and 3.10).

For linear motions we develop a simple numerical method for estimating depth and surface curvatures from a minimum of three discrete views, by determining the *osculating* circle in each epipolar plane. The error and sensitivity analysis is greatly simplified with this formulation. Of course this introduces a tradeoff between the scale at which curvature is measured (truncation error) and measurement error. We are no longer computing surface curvature at a point but bounds on surface curvature. However the computation allows the use of longer "stereo baselines" and is less sensitive to edge localisation.

## Numerical method for depth and curvature estimation

Consider three views taken at times $t_0$, $t_1$, and $t_2$ from camera positions $\mathbf{v}(t_0)$, $\mathbf{v}(t_1)$ and $\mathbf{v}(t_2)$ respectively (figure 3.9). Let us select a point on an image contour in the first view, say $\mathbf{p}(s_0, t_0)$. For linear motion and epipolar parameterisation the *corresponding* ray directions and the contact point locus, $\mathbf{r}(s_0, t)$, lie in a plane – the epipolar plane. Analogous to stereo matching *corresponding* features are found by searching along epipolar lines in the subsequent views. The three rays are tangents to $\mathbf{r}(s_0, t)$. They do not, in general, define a unique curve (figure 3.10). They may, however, constrain its curvature. By assuming that the curvature of the curve $\mathbf{r}(s_0, t)$ is locally constant it can be approximated as part of a circle (in the limit the *osculating circle*) of radius R (the reciprocal of curvature) and with centre at $\mathbf{p}_0$ such that (figure 3.10):

$$\mathbf{r}(s_0, t) = \mathbf{p}_0 + R\mathbf{N}(s_0, t) \qquad (3.21)$$

where $\mathbf{N}$ is the Frenet–Serret curve normal in each view. $\mathbf{N}$ is perpendicular to the ray direction and, in the case of epipolar parameterisation, lies in the epipolar plane (the *osculating* plane). It is defined by two components in this plane.

Since the rays $\mathbf{p}(s_0, t)$ are tangent to the curve we can express (3.21) in terms of image measurables, $\mathbf{N}(s_0, t)$, and unknown quantities $\mathbf{p}_0$ and $R$:

$$(\mathbf{r}(s_0, t) - \mathbf{v}(t)).\mathbf{N}(s_0, t) = 0$$
$$(\mathbf{p}_0 + R\mathbf{N}(s_0, t) - \mathbf{v}(t)).\mathbf{N}(s_0, t) = 0. \qquad (3.22)$$

These quantities can be uniquely determined from measurements in three distinct views. For convenience we use subscripts to label the measurements made for each view (discrete time).

$$
\begin{aligned}
\mathbf{p}_0.\mathbf{N}_0 + R &= \mathbf{v}_1.\mathbf{N}_0 \\
\mathbf{p}_0.\mathbf{N}_1 + R &= \mathbf{v}_2.\mathbf{N}_1 \\
\mathbf{p}_0.\mathbf{N}_2 + R &= \mathbf{v}_3.\mathbf{N}_2.
\end{aligned}
\tag{3.23}
$$

Equations (3.23) are linear equations in three unknowns (two components of $\mathbf{p}_0$ in the epipolar plane and the radius of curvature, $R$) and can be solved by standard techniques. If more than three views are processed the over-determined system of linear equations of the form of (3.23) can be solved by least squares.

For a general motion in $R^3$ the camera centres will not be collinear and the epipolar structure will change continuously. The three rays will not in general lie in a common epipolar plane (the osculating plane) since the space curve $\mathbf{r}(s_0, t)$ now has torsion. The first two viewpoints, however, define an epipolar plane which we assume is the osculating plane of $\mathbf{r}(s_0, t)$. Projecting the third ray on to this plane allows us to recover an approximation for the osculating circle and hence $R$, which is correct in the limit as the spacing between viewpoints becomes infinitesimal. This approximation is used by Vaillant and Faugeras [203, 204] in estimating surface shape from trinocular stereo with cameras whose optical centres are not collinear.

### Experimental results – curvature from three discrete views

The three views shown in figure 3.9 are from a sequence of a scene taken from a camera mounted on a moving robot-arm whose position and orientation have been accurately calibrated from visual data for each viewpoint [195]. The image contours are tracked automatically (figure 3.4) and equations (3.23) are used to estimate the radius of curvature of the *epipolar section*, R, for a point A on an extremal boundary of the vase. The method is repeated for a point which is not on an extremal boundary but is on a nearby surface marking, B. As before this is a degenerate case of the parameterisation.

The radius of curvature at A was estimated as $42 \pm 15$mm. It was measured using calipers as $45 \pm 2$mm. For the marking, B, the radius of curvature was estimated as $3 \pm 15$mm. The estimated curvatures agree with the actual curvatures. However, the results are very sensitive to perturbations in the assumed values of the motion and to errors in image contour localisation (figure 3.11).

## 3.4   Error and sensitivity analysis

The estimate of curvature is affected by errors in image localisation and uncertainties in ego-motion calibration in a non-linear way. The effect of small errors in the assumed ego-motion is computed below.

The radius of curvature $R$ can be expressed as a function $g$ of $m$ variables $w_i$:

$$R = g(w_1, w_2, \ldots w_m) \tag{3.24}$$

where typically $w_i$ will include image positions $(\mathbf{q}(s_0, t_0), \mathbf{q}(s_0, t_1), \mathbf{q}(s_0, t_2))$; camera orientations $(\mathbf{R}(t_0), \mathbf{R}(t_1), \mathbf{R}(t_2))$; camera positions $(\mathbf{v}(t_0), \mathbf{v}(t_1), \mathbf{v}(t_2))$; and the intrinsic camera parameters. The effect on the estimate of the radius of curvature, $\delta R$, of small systematic errors or biases, $\delta w_i$, can be easily computed, by first-order perturbation analysis.

$$\delta R = \sum_i \frac{\partial g}{\partial w_i} \delta w_i. \tag{3.25}$$

The propagation of uncertainties in the measurements to uncertainties of the estimates can be similarly derived. Let the variance $\sigma_{w_i}^2$ represent the uncertainty of the measurement $w_i$. We can propagate the effect of these uncertainties to compute the uncertainty in the estimate of $R$ [69]. The simplest case is to consider the error sources to be statistically independent and uncorrelated. The uncertainty in $R$ is then

$$\sigma_R^2 = \sum_i \left(\frac{\partial g}{\partial w_i}\right)^2 \sigma_{w_i}^2. \tag{3.26}$$

These expressions will now be used to analyse the sensitivity to viewer ego-motion of absolute and parallax-based measurements of surface curvature. They will be used in the next section in the hypothesis test to determine whether the image contour is the projection of a fixed feature or is extremal. That is, to test whether the radius of curvature is zero or not.

## Experimental results – sensitivity analysis

The previous section showed that the visual motion of apparent contours can be used to estimate surface curvatures of a useful accuracy if the viewer ego-motion is known. However, the estimate of curvature is very sensitive to perturbations in the motion parameters. The effect of small errors in the assumed ego-motion – position and orientation of the camera – is given by (3.25) and are plotted in figure 3.12a and 3.12b (curves labelled I). Accuracies of 1 part in 1000 in the measurement of ego-motion are essential for surface curvature estimation.

Parallax based methods measuring surface curvature are in principle based on measuring the relative image motion of nearby points on different contours (2.59). In practice this is equivalent (equation(2.57)) to computing the difference of radii of curvature at the two points, say A and B (figure 3.9). The radius of

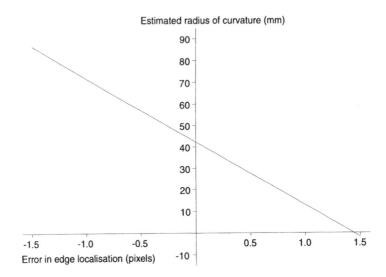

Figure 3.11: Sensitivity of curvature estimate to errors in image contour localisation.

curvature measured at a surface marking is determined by errors in image measurement and ego-motion. (For a precisely known viewer motion and for exact
contour localisation the radius of curvature would be zero at a fixed feature.)
It can be used as a reference point to subtract the global additive errors due to
imprecise motion when estimating the curvature at the point on the extremal
boundary. Figures 3.12a and 3.12b (curves labelled II) show that the sensitivity
of the relative inverse curvature, $\Delta R$, to error in position and rotation computed
between points A and B (two nearby points at similar depths) is reduced by an
order of magnitude. This is a striking decrease in sensitivity even though the
features do not coincide exactly as the theory required.

## 3.5 Detecting extremal boundaries and recovering surface shape

### 3.5.1 Discriminating between fixed features and extremal boundaries

The magnitude of $R$ can be used to determine whether a point on an image
contour lies on an apparent contour or on the projection of a fixed surface feature
such as a crease, shadow or surface marking.

    With noisy image measurements or poorly calibrated motion we must test

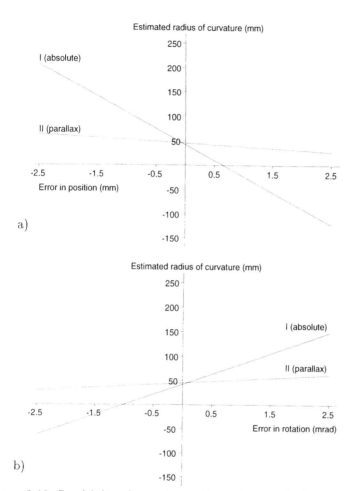

Figure 3.12: Sensitivity of curvature estimated from absolute measurements and parallax to errors in motion.

(a) The radius of curvature $(R = 1/\kappa^t)$ for a point on the extremal boundary (A) is plotted as a function of errors in the camera position (a) and orientation (b). Curvature estimation is highly sensitive to errors in egomotion. Curve I shows that a perturbation of 1mm in position (in a translation of 100mm) produces an error of 155% in the estimated radius of curvature. A perturbation of 1mrad in rotation about an axis defined by the epipolar plane (in a total rotation of 200mrad) produces an error of 100%.

(b) However, if parallax-based measurements are used the estimation of curvature is much more robust to errors in egomotion. Curve II shows the difference in radii of curvature between a point on the extremal boundary (A) and the nearby surface marking (B) plotted against error in the position (a) and orientation (b). The sensitivity is reduced by an order of magnitude, to 19% per mm error and 12% per mrad error respectively.

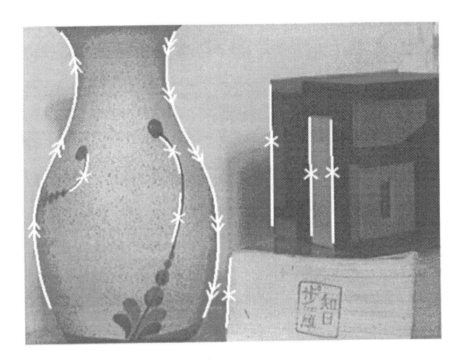

Figure 3.13: Detecting and labelling extremal boundaries.

*The magnitude of the radius of curvature ($1/\kappa^t$, computed from 3 views) can be used to classify image curves as either the projection of extremal boundaries or fixed features (surface markings, occluding edges or orientation discontinuities). The sign of $\kappa^t$ determines on which side of the image contour lies the surface. NOTE: a $\times$ label indicates a fixed feature. A $\gg\!\!-$ label indicates an apparent contour. The surface lies to the right as one moves in the direction of the twin arrows [141]. The sign of Gaussian curvature can then be inferred directly from the sign of the curvature of the apparent contour.*

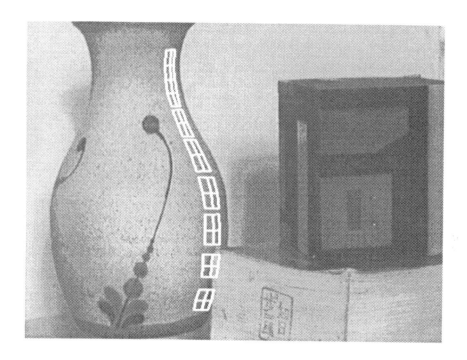

Figure 3.14: Recovery of surface strip in vicinity of extremal boundary.
*From a minimum of three views of a curved surface it is possible to recover the
3D geometry of the surface in the vicinity of extremal boundary. The surface
is recovered as a family of s-parameter curves – the contour generators – and
t-parameter curves – portions of the osculating circles measured in each epipolar
plane. The strip is shown projected into the image of the scene from a different
viewpoint*

Figure 3.15: Reconstructed surface.

*Reconstructed surface obtained by extrapolation of computed surface curvatures in the vicinity of the extremal boundary (A) of the vase, shown here from a new viewpoint.*

by error analysis the hypothesis that $R$ is not equal to zero for an extremal boundary. We have seen how to compute the effects of small errors in image measurement, and ego-motion. These are conveniently represented by the *co-variance* of the estimated curvature. The estimate of the radius of curvature and its uncertainty is then used to test the hypothesis of an extremal boundary. In particular if we assume that the error in the estimate of the radius has a Normal distribution (as an approximation to the Student–t distribution [178]), the image contour is assumed to be the projection of a fixed feature (within a confidence interval of 95%) if:

$$-1.96\sigma_R < R < 1.96\sigma_R. \tag{3.27}$$

Using *absolute measurements*, however, the discrimination between fixed and extremal features is limited by the uncertainties in robot motion. For the image sequence of figure 3.9 it is only possible to discriminate between fixed features and points on extremal boundaries with inverse curvatures greater than 15mm. High curvature points ($R < 1.96\sigma_R$) cannot be distinguished from fixed features and will be incorrectly labelled.

By using *relative measurements* the discrimination is greatly improved and is limited by the finite separation between the points as predicted by (2.62). For the example of figure 3.9 this limit corresponds to a relative curvature of approximately 3mm. This, however, requires that we have available a fixed nearby reference point.

Suppose now that no known surface feature has been identified in advance. Can the robust relative measurements be made to bootstrap themselves without an independent surface reference? It is possible by relative (two-point) curvature measurements obtained for a small set of nearby points to determine pairs which are fixed features. They will have zero relative radii of curvature. Once a fixed feature is detected it can act as stable reference for estimating the curvature at extremal boundaries.

In detecting an apparent contour we have also determined on which side the surface lies and so can compute the sign of Gaussian curvature from the curvature of the image contour. Figure 4.13 shows a selected number of contours which have been automatically tracked and are correctly labelled by testing for the sign and magnitude of $R$.

### 3.5.2 Reconstruction of surfaces

In the vicinity of the extremal boundary we can recover the two families of para-metric curves. These constitute a *conjugate* grid of surface curves: *s*-parameter curves (three extremal contour generators from the different viewpoints) and *t*-parameter curves (the intersection of a pencil of epipolar planes defined by the

first two viewpoints and the surface). The recovered strip of surface is shown in figure 3.14 projected into the image from a fourth viewpoint. The reconstructed surface obtained by extrapolation of the computed surface curvatures at the extremal boundary A of the vase is shown from a new viewpoint in figure 3.15.

## 3.6   Real-time experiments exploiting visually derived shape information

Preliminary results for three tasks involving a CCD camera mounted on the wrist joint of a robot-arm are now described. In the previous section it was shown how a camera mounted on a robot manipulator can be moved around an unknown object, detect the extremal contours and incrementally construct a 3D geometric model of the object from their deformation (figure 3.15).

In a second task it is shown how surface curvature is used to aid path planning around curved objects. The camera makes deliberate movements and tracks image contours. Estimates of distance and curvature are used to map out a safe, obstacle-free path around the object. Successful inference and reasoning about 3D shape are demonstrated by executing the motion.

In a third task the power of robust parallax-based estimates of surface curvature is demonstrated in an experiment in which the relative motion of two nearby contours is used to refine the estimates of surface curvatures of an unknown object. This information is used to plan an appropriate grasping strategy and then grasp and manipulate the object.

### 3.6.1   Visual navigation around curved objects

In this section results are presented showing how a moving robot manipulator can exploit the visually derived 3D shape information to plan a smooth, safe path around an obstacle placed in its path. The scenario of this work is that the start and goal position for a mobile camera are fixed and the robot is instructed to reach the goal from the start position, skirting around any curved obstacles that would be encountered on a straight line path from the current position to the goal. The camera first localises an apparent contour and makes a small sideways motion to generate visual motion. This allows it to compute the distance to the contour generator and more importantly the curvature of the visible surface in the epipolar plane. A safe path around the curved object is then planned by extrapolating the computed curvatures with a correction to allow for the uncertainty so as to ensure safe clearances. The robot then steers the camera around the obstacle with a clearance of a few millimetres. Examples running with single image contours are shown in figures 3.16 and 3.17.

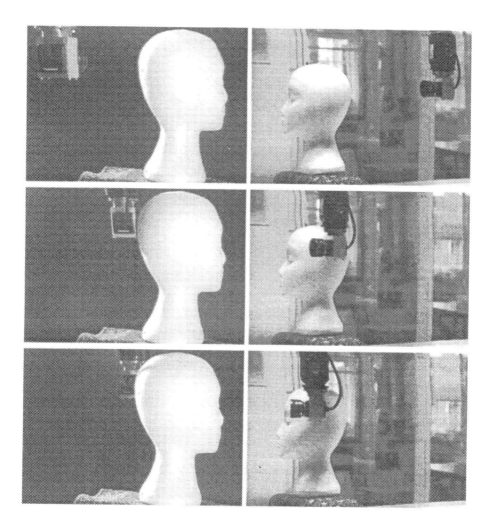

Figure 3.16: Visually guided navigation around curved obstacles.

*The visual motion of an apparent contour under known viewer motion is used to estimate the position, orientation, and surface curvature of the visible surface. In addition to this quantitative information the visual motion of the apparent contour can also determine which side of the contour is free space. This qualitative and quantitative information is used to map out a safe path around the unmodelled obstacle. The sequence of images shows the robot manipulator's safe execution of the planned path, seen from two viewpoints.*

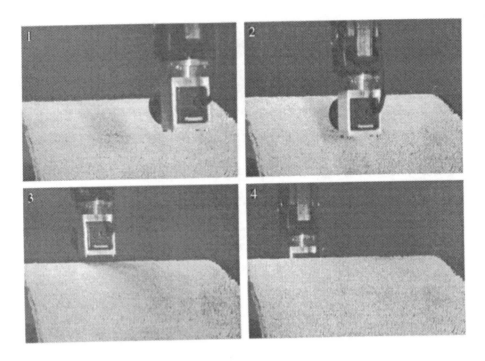

Figure 3.17: Visually guided navigation over undulating surface.

*After detecting a horizontal image contour, the image motion due to a small local vertical viewer motion is used to estimate the distance to the contour generator. A larger extended motion is then used to estimate the surface curvature at the contour generator. This is used to map out and execute a safe path over the obstacle, shown in this sequence of images.*

The path planning algorithms for navigating around curved surfaces are further developed in Blake et al. [24]. They show that minimal paths are smooth splines composed of *geodesics* [67] and straight lines in free space. Computation of the geodesics, in general, requires the complete 3D surface. In the case where geometric information is imperfect, in that surface shape is not known a priori, they show that it is possible to compute a helical approximation to the sought geodesic, based only on the visible part of the surface near the extremal boundary. The information required for the helical approximation can be computed directly from the deformation of the apparent contour.

## 3.6.2   Manipulation of curved objects

Surface curvature recovered directly from the deformation of the apparent contour (instead of dense depth maps) yields useful information for path planning. This information is also important for grasping curved objects.

Reliable estimates of surface curvature can be used to determine grasping points. Figure 3.18 shows an example of a scene with a vase placed approximately 1m away from a robot manipulator and suction gripper. Estimates of surface curvature at the extremal boundary are used to position a suction gripper for manipulation. The robot initialises a snake which localises a nearby high contrast edge. In the example shown the snake initially finds the edge of the cardboard box (B). The robot then makes a small local motion of a few centimetres to estimate the depth of the feature. It uses this information so that it can then track the contrast edge over a larger baseline while fixating (keeping the edge in the centre of the image). Before executing the larger motion of 15cm the first snake (parent) spawns a child snake which finds a second nearby edge (A). The two edges are then tracked together, allowing the accurate estimation of surface curvature by reducing the sensitivity to robot "wobble" and systematic errors in the robot motion (see figures 3.7 and 3.8). The estimates of curvature are accurate to a few millimetres. This is in contrast to estimates of curvature based on the absolute motion of an apparent contour which deliver curvature estimates which are only correct to the nearest centimetre. The extrapolation of these surface curvatures allows the the robot to plan a grasping position which is then successfully executed (figure 3.18).

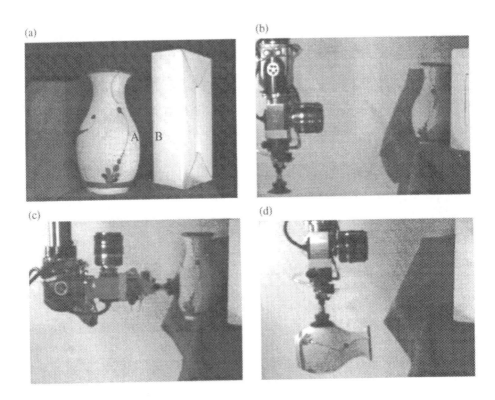

Figure 3.18: Visually guided manipulation of piecewise curved objects.

*The manipulation of curved objects requires precise 3D shape (curvature) information. The accuracy of measurements of surface curvature based on the deformation of a single apparent contour is limited by uncertainty in the viewer motion. The effect of errors in viewer motion is greatly reduced and the accuracy of surface curvature estimates consequently greatly improved by using the relative motion of nearby image contours. In the example shown the relative motion between the image of the projection of the crease of the box and the apparent contour of the vase is used to estimate surface curvature to the nearest 5mm (of a measurement of 40mm) and the contour generator position to the nearest 1mm (at a distance of 1m). This information is used to guide the manipulator and suction gripper to a convenient location on the surface of the vase for manipulation.*

# Chapter 4

# Qualitative Shape from Images of Surface Curves

## 4.1  Introduction

Imagine we have several views of a curve lying on a surface. If the motion between the views and the camera calibration are known then in principle it is possible to reconstruct this space curve from its projections. It is also possible in principle to determine the curve's tangent and curvature. In practice this might require the precise calibration of the epipolar geometry and sub-pixel accuracy for edge localisation and/or integrating information over many views in order to reduce discretisation errors.

However, even if perfect reconstruction could be achieved, the end result would only be a space curve. This delimits the surface, but places only a weak constraint on the surface orientation and shape along the curve (via the visibility and tangent constraints - see later). Ideally, rather than simply a space curve we would like a *surface strip* [122] along which we know the surface orientation. Better still would be knowledge of how the surface normal varied not only along the curve but also in arbitrary directions away from the curve. This determines the principal curvatures and direction of the principal axes along the strip. This information is sufficient to completely specify the surface shape locally. Knowledge of this type helps to infer surface behaviour away from the curves, and thus enables grouping of the curves into coherent surfaces.

For certain surface curves and tracked points the information content is not so bleak. It was shown in Chapter 2 that the surface normal is known along the apparent contour (the image of the points where the viewing direction lies in the tangent plane) [17]. Further, the curvature of the apparent contour in a single view determines the sign of the Gaussian curvature of the surface projecting to the contour [120, 36]. From the deformation of the apparent contour under viewer motion a surface patch (first and second fundamental forms) can be recovered [85, 27]. The deformation of image curves due to viewer motion, also allows us to *discriminate* the image of surface curves from apparent contours. A self-shadow (where the illuminant direction lies in the tangent plane) can be

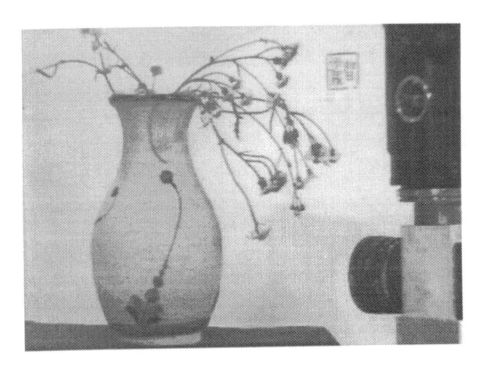

Figure 4.1: Qualitative shape from the deformation of image curves.

*A single CCD camera mounted on the wrist joint of a 5-axis Adept 1 SCARA arm (shown on right) is used to recover qualitative aspects of the geometry of visible surfaces from a sequence of views of surface curves.*

exploited in a similar manner if the illuminant position is known [122]. Tracking specular points [220] gives a surface strip along which the surface normal is known.

In this chapter we analyse the images of surface curves (contour generators which arise because of internal surface markings or illumination effects) and investigate the surface geometric information available from the temporal evolution of the image under viewer motion (figure 4.1).

Surface curves have three advantages over isolated surface markings:

1. Sampling – Isolated texture only "samples" the surface at isolated points – the surface could have any shape in between the points. Conversely, a surface curve conveys information, at a particular scale, throughout its path.

2. Curves, unlike points, have well-defined tangents which constrain surface orientation.

3. Technological – There are now available reliable, accurate edge detectors which localise surface markings to sub-pixel accuracy [48]. The technology for isolated point detection is not at such an advanced stage. Furthermore, snakes [118] are ideally suited to tracking curves through a sequence of images, and thus measuring the curve deformation (Chapter 3).

This chapter is divided into three parts. First, in section 4.2, the geometry of space curves is reviewed and related to the perspective image. In particular, a simple expression for the curvature of the image contour is derived. Second, in section 4.3, the information available from the deformation of the image curve under viewer motion is investigated, making explicit the constraints that this imposes on the geometry of the space curve. Third, in section 4.4, the aspects of the differential geometry of the surface that can be gleaned by knowing that the curve lies on the surface are discussed.

The main contribution concerns the recovery of aspects of qualitative shape. That is, information that can be recovered efficiently and robustly, without requiring exact knowledge of viewer motion or accurate image measurements. The description is, however, incomplete. It is shown that *visibility* of points on the curve places a weak constraint on the surface normal. This constraint is tightened by including the restriction imposed by the surface curve's tangent. Furthermore, certain 'events' (inflections, transverse curve crossings) are richer still in geometric information. In particular it is shown that tracking image curve inflections determines the sign of the normal curvature in the direction of the surface curve's tangent vector. This is a generalisation to surface curves of Weinshall's [212] result for surface texture. Examples are included for real image sequences.

In addition to the information that surface curves provide about surface shape, the deformation also provides constraints on the viewer (or object) motion. This approach was introduced by Faugeras [71] and is developed in section 4.5.

## 4.2   The perspective projection of space curves

### 4.2.1   Review of space curve geometry

Consider a point $P$ on a regular differentiable space curve $\mathbf{r}(s)$ in $R^3$ (figure 4.2a). The local geometry of the curve is uniquely determined in the neighbourhood of $P$ by the basis of unit vectors $\{\mathbf{T}, \mathbf{N}, \mathbf{B}\}$, the curvature, $\kappa$, and torsion, $\tau$, of the space curve [67]. For an arbitrary parameterisation of the curve, $\mathbf{r}(s)$, these quantities are defined in terms of the derivatives (up to third order) of the curve with respect to the parameter $s$. The first-order derivative ("velocity") is used to define the tangent to the space curve, $\mathbf{T}$, a unit vector given by

$$\mathbf{T} = \frac{\mathbf{r}_s}{|\mathbf{r}_s|}. \tag{4.1}$$

The second-order derivative – in particular the component perpendicular to the tangent ("centripetal acceleration") – is used to define the curvature, $\kappa$ (the magnitude) and the curve normal, $\mathbf{N}$ (the direction):

$$\kappa\mathbf{N} = \frac{(\mathbf{T} \wedge \mathbf{r}_{ss}) \wedge \mathbf{T}}{|\mathbf{r}_s|^2}. \tag{4.2}$$

The plane spanned by $\mathbf{T}$ and $\mathbf{N}$ is called the *osculating plane*. This is the plane which $\mathbf{r}(s)$ is closest to lying in (and does lie in if the curve has no torsion). These two vectors define a natural frame for describing the geometry of the space curve. A third vector, the binormal $\mathbf{B}$, is chosen to form a right-handed set:

$$\mathbf{B} = \mathbf{T} \wedge \mathbf{N}. \tag{4.3}$$

This leaves only the torsion of the curve, defined in terms of deviation of the curve out of the osculating plane:

$$\tau = \frac{\mathbf{r}_{sss}.\mathbf{B}}{\kappa\,|\mathbf{r}_s|^3}. \tag{4.4}$$

The relationship between these quantities and their derivatives for movements along the curve can be conveniently packaged by the Frenet–Serret equations [67] which for an arbitrary parameterisation are given by:

$$\mathbf{T}_s = |\mathbf{r}_s|\kappa\mathbf{N} \tag{4.5}$$

$$\mathbf{N}_s = |\mathbf{r}_s|(-\kappa\mathbf{T} + \tau\mathbf{B}) \tag{4.6}$$

$$\mathbf{B}_s = -|\mathbf{r}_s|\tau\mathbf{N}. \tag{4.7}$$

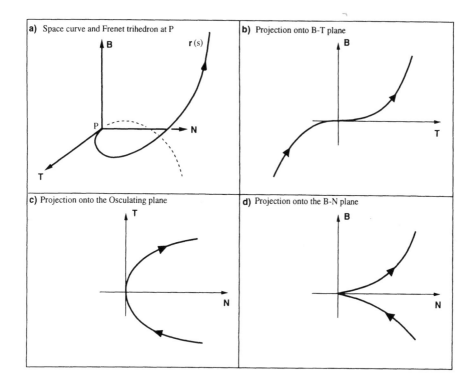

Figure 4.2: Space curve geometry and local forms of its projection.

*The local geometry of a space curve can be completely specified by the Frenet trihedron of vectors $\{\mathbf{T}, \mathbf{N}, \mathbf{B}\}$, the curvature, $\kappa$, and torsion, $\tau$, of the curve. Projection of the space curve onto planes perpendicular to these vectors ( the local canonical forms [67]) provides insight into how the apparent shape of a space curve changes with different viewpoints.*

The influence of curvature and torsion on the shape of a curve are clearly demonstrated in the Taylor series expansion by arc length about a point $u_0$ on the curve.

$$\mathbf{r}(u) = \mathbf{r}(u_0) + u\mathbf{r}_u(u_0) + \frac{u^2}{2}\mathbf{r}_{uu}(u_0) + \frac{u^3}{6}\mathbf{r}_{uuu}(u_0) \quad \ldots \qquad (4.8)$$

where $u$ is an arc length parameter of the curve. An approximation for the curve with the lowest order in $u$ along each basis vector is given by [122]:

$$\mathbf{r}(u) = \mathbf{r}(u_0) + (u + \ldots)\mathbf{T} + (\frac{u^2}{2}\ldots)\kappa\mathbf{N} + (\frac{u^3}{6} + \ldots)\kappa\tau\mathbf{B} \qquad (4.9)$$

The zero-order term is simply the fiducial point itself; the first-order term is a straight line along the tangent direction; the second-order term is a parabolic arc in the osculating plane; and the third-order term describes the deviation from the osculating plane. Projection on to planes perpendicular to $\mathbf{T}, \mathbf{N}, \mathbf{B}$ give the local forms shown in figure 4.2. It is easy to see from (4.9) that the orthographic projection on to the $\mathbf{T} - \mathbf{N}$ plane (osculating plane) is just a parabolic arc; on the $\mathbf{T} - \mathbf{B}$ plane you see an inflection; and the projection on the $\mathbf{N} - \mathbf{B}$ plane is a *cusped* curve. If $\kappa$ or $\tau$ are zero then higher order terms are important and the local forms must be modified. These local forms provide some insight into how the apparent shape of a space curve changes with different viewpoint. The exact relationship between the space curve geometry and its image under perspective projection will now be derived.

## 4.2.2  Spherical camera notation

As in Chapter 2, consider perspective projection on to a sphere of unit radius. The advantage of this approach is that formulae under perspective are often as simple as (or identical to) those under orthographic projection [149].

The image of a world point, $P$, with position vector, $\mathbf{r}(s)$, is a unit vector $\mathbf{p}(s,t)$ such that [1]

$$\mathbf{r}(s) = \mathbf{v}(t) + \lambda(s,t)\mathbf{p}(s,t) \qquad (4.10)$$

where $s$ is a parameter along the image curve; $t$ is chosen to index the view (corresponding to time or viewer position) $\lambda(s,t)$ is the distance along the ray to $P$; and $\mathbf{v}(t)$ is the viewer position (centre of spherical pin-hole camera) at time $t$ (figure 4.3). A moving observer at position $\mathbf{v}(t)$ sees a family of views of the curve indexed by time, $\mathbf{q}(s,t)$ (figure 4.4).

---

[1]The space curve $\mathbf{r}(s)$ is fixed on the surface and is view independent. This is the only difference between (4.10) and (2.10).

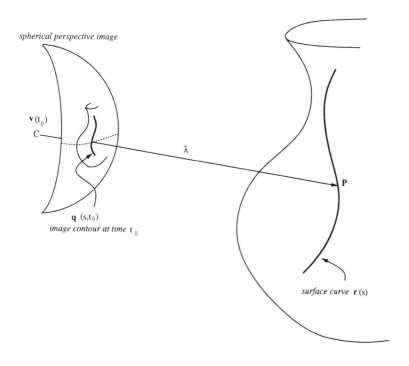

spherical perspective image

$\mathbf{v}\,(t_0)$

C

$\lambda$

P

$\mathbf{q}\,(s,t_0)$

image contour at time $t_0$

surface curve $\mathbf{r}\,(s)$

Figure 4.3: Viewing and surface geometry.

The image defines the direction of a ray, (unit vector $\mathbf{p}$) to a point, P, on a surface curve, $\mathbf{r}(s)$. The distance from the viewer (centre of projection sphere) to P is $\lambda$.

### 4.2.3   Relating image and space curve geometry

Equation (4.10) gives the relationship between a point on the curve $\mathbf{r}(s)$, and its spherical perspective projection, $\mathbf{p}(s,t)$, for a view indexed by time $t$. It can be used to relate the space curve geometry $(\mathbf{T}, \mathbf{N}, \mathbf{B}, \kappa, \tau)$ to the image and viewing geometry. The relationship between the orientation of the curve and its image tangent and the curvature of the space curve and its projection are now derived.

**Image curve tangent and normal**

At the projection of $P$, the tangent to the spherical image curve, $\mathbf{t}^p$, is related to the space curve tangent $\mathbf{T}$ and the viewing geometry by:

$$\mathbf{t}^p = \frac{\mathbf{T} - (\mathbf{p}.\mathbf{T})\mathbf{p}}{(1 - (\mathbf{p}.\mathbf{T})^2)^{1/2}}. \tag{4.11}$$

**Derivation 4.1** *Differentiating (4.10) with respect to $s$,*

$$\mathbf{r}_s = \lambda_s \mathbf{p} + \lambda \mathbf{p}_s \tag{4.12}$$

*and rearranging we derive the following relationships:*

$$\mathbf{p}_s = \frac{\mathbf{p} \wedge (\mathbf{r}_s \wedge \mathbf{p})}{\lambda} \tag{4.13}$$

$$|\mathbf{p}_s| = \frac{|\mathbf{r}_s|}{\lambda} \left( 1 - \left( \mathbf{p} . \frac{\mathbf{r}_s}{|\mathbf{r}_s|} \right)^2 \right)^{1/2}. \tag{4.14}$$

*Note that the mapping from space curve to the image contour is singular (degenerate) when the ray and curve tangent are aligned. The tangent to the space curve projects to a point in the image and a cusp is generated in the image contour.*

*By expressing (4.13) in terms of unit tangent vectors, $\mathbf{t}^p$ and $\mathbf{T}$:*

$$\begin{aligned} \mathbf{t}^p &= \frac{\mathbf{p}_s}{|\mathbf{p}_s|} \\ &= \frac{\mathbf{p} \wedge (\mathbf{T} \wedge \mathbf{p})}{(1 - (\mathbf{p}.\mathbf{T})^2)^{1/2}}. \end{aligned}$$

The direction of the ray, $\mathbf{p}$, and the image curve tangent $\mathbf{t}^p$ determine the orientation of the image curve normal $\mathbf{n}^p$:

$$\mathbf{n}^p = \mathbf{p} \wedge \mathbf{t}^p. \tag{4.15}$$

Note, this is not the same as the projection of the surface normal, $\mathbf{n}$. However, it is shown below that the image curve normal $\mathbf{n}^p$ constrains the surface normal.

## Curvature of projection

A simple relationship between the shape of the image and space curves and the viewing geometry is now investigated. In particular, the relationship between the curvature of the image curve, $\kappa^p$ (defined as the *geodesic* curvature [2] of the spherical curve, $\mathbf{p}(s, t)$) is derived:

$$\kappa^p = \frac{\mathbf{p}_{ss}.\mathbf{n}^p}{|\mathbf{p}_s|^2} \qquad (4.16)$$

and the space curve curvature, $\kappa$:

$$\kappa^p = \lambda\kappa \frac{[\mathbf{p}, \mathbf{T}, \mathbf{N}]}{[1 - (\mathbf{p}.\mathbf{T})^2]^{3/2}}, \qquad (4.17)$$

where $[\mathbf{p}, \mathbf{T}, \mathbf{N}]$ represents the triple scalar product. The numerator depends on the angle between the ray and the osculating plane. The denominator depends on the angle between the ray and the curve tangent.

**Derivation 4.2** *Differentiating (4.12) with respect to s and collecting the components parallel to the image curve normal gives*

$$\mathbf{p}_{ss}.\mathbf{n}^p = \frac{\mathbf{r}_{ss}.\mathbf{n}^p}{\lambda}. \qquad (4.18)$$

*Substituting this and (4.14) into the expression for the curvature of the image curve (4.16)*

$$\kappa^p = \frac{\mathbf{r}_{ss}.\mathbf{n}^p}{\lambda|\mathbf{p}_s|^2} \qquad (4.19)$$

$$= \lambda\kappa \frac{\mathbf{N}.\mathbf{n}^p}{(1 - (\mathbf{p}.\mathbf{T})^2)}. \qquad (4.20)$$

*Substituting (4.15) and (4.11) for $\mathbf{n}^p$:*

$$\kappa^p = \lambda\kappa \frac{\mathbf{N}.(\mathbf{p} \wedge \mathbf{T})}{(1 - (\mathbf{p}.\mathbf{T})^2)^{3/2}} \qquad (4.21)$$

$$= \lambda\kappa \frac{\mathbf{B}.\mathbf{p}}{(1 - (\mathbf{p}.\mathbf{T})^2)^{3/2}}. \qquad (4.22)$$

A similar result is described in [122]. Under orthographic projection the expression is the same apart from the scaling factor of $\lambda$. As expected, the image curvature scales linearly with distance and is proportional to the space curve curvature $\kappa$. More importantly, the sign of the curvature of the projection depends on which side of the osculating plane the ray, $\mathbf{p}$ lies, i.e. the sign of the

---

[2] The *geodesic* curvature of a space curve has a well-defined sign. It is, of course meaningless to refer to the sign of curvature of a space curve.

scalar product $\mathbf{B}.\mathbf{p}$. That is easily seen to be true by viewing a curve drawn on a sheet of paper from both sides. The case in which the vantage point is in the osculating plane corresponds to a zero of curvature in the projection. From (4.17) (see also [205]) the projected curvature will be zero if and only if:

1. $\kappa = 0$

   The curvature of the space curve is zero. This does not occur for generic curves [41]. Although the projected curvature is zero, this may be a zero *touching* rather than a zero crossing of curvature.

2. $[\mathbf{p}, \mathbf{T}, \mathbf{N}] = 0$ with $\mathbf{p}.\mathbf{T} \neq 0$

   The view direction lies in the osculating plane, but not along the tangent to the curve (if the curve is projected along the tangent the image is, in general, a cusp). Provided the torsion is not zero, $\mathbf{r}(s)$ crosses its osculating plane, seen in the image as a zero crossing.

Inflections will occur generically in any view of a curve, but cusps only become generic in a one-parameter family of views [41]. [3] Inflections in image curves are therefore more likely to be consequences of the viewing geometry (condition 2 above) than zeros of the space curve curvature (condition 1). Contrary to popular opinion [205] the power of inflections of image curves as invariants of perspective projection of space curves is therefore limited.

## 4.3   Deformation due to viewer movements

As the viewer moves the image of $\mathbf{r}(s)$ will deform. The deformation is characterised by a change in image position (image velocities), a change in image curve orientation and a change in the curvature of the projection. Below we derive expressions relating the deformation of the image curve to the space curve geometry and then show how to recover the latter from simple measurements on the spatio-temporal image.

Note that for a moving observer the viewer (camera) co-ordinate is continuously changing with respect to the fixed co-ordinate system used to describe $R^3$ (see section 2.2.4). The relationship between temporal derivatives of measurements made in the camera co-ordinate system and those made in the reference

---

[3] An informal way to see this is to consider orthographic projection with the view direction defining a point $\mathbf{p}$ on the Gaussian sphere. The tangent at each point on the space curve also defines a point on the Gaussian sphere, and so $\mathbf{T}_G(s)$ traces a curve. For a cusp, $\mathbf{p}$ must lie on $\mathbf{T}_G(s)$ and this will not occur in general. However, a one parameter family of views $\mathbf{p}(t)$ also defines a curve on the Gaussian sphere. Provided these cross (transversely) the intersection will be stable to perturbations in $\mathbf{r}(s)$ (and hence $\mathbf{T}_G(s)$) and $\mathbf{p}(t)$. A similar argument establishes the inflection case.

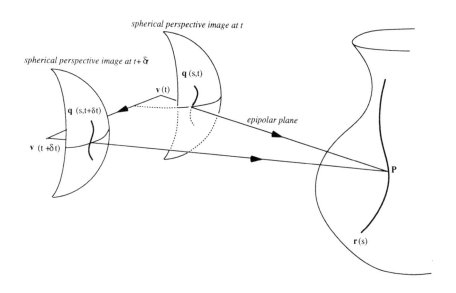

Figure 4.4: Epipolar geometry.

A moving observer at position $\mathbf{v}(t)$ sees a one-parameter family of image curves, $\mathbf{p}(s,t)$ – the spherical perspective projections of a space curve, $\mathbf{r}(s)$, indexed by time. Knowledge of the viewer's motion (camera centre and orientation) is sufficient to determine the corresponding image point (and hence the direction of the ray) in successive images. These are found by searching along epipolar great-circles. The space curve can then be recovered by triangulation of the viewer positions and the ray directions from a minimum of two views.

frame is obtained by differentiating (2.12) and (2.13). In particular the temporal derivative of the ray, the image curve tangent and the image curve normal, $\{\mathbf{p}, \mathbf{t}^p, \mathbf{n}^p\}$, are related to temporal derivatives of the image curve measured in the viewer co-ordinate system, $\{\mathbf{q}, \tilde{\mathbf{t}}^p, \tilde{\mathbf{n}}^p\}$, by

$$
\begin{align}
\mathbf{p}_t &= \mathbf{q}_t + \Omega(t) \wedge \mathbf{q} \tag{4.23}\\
\mathbf{t}_t^p &= \tilde{\mathbf{t}}_t^p + \Omega(t) \wedge \tilde{\mathbf{t}}^p \tag{4.24}\\
\mathbf{n}_t^p &= \tilde{\mathbf{n}}_t^p + \Omega(t) \wedge \tilde{\mathbf{n}}^p. \tag{4.25}
\end{align}
$$

For a static space curve (not an extremal boundary of a curved surface) a point $P$ on the curve, $\mathbf{r}(s)$, does not change with time:

$$\mathbf{r}_t = 0. \tag{4.26}$$

This can be used to derive the relationship between the images of the point $P$ in the sequence of views. Differentiating (4.10) with respect to $t$ and substituting the condition (4.26) gives an infinitesimal analogue of the epipolar constraint in which the ray is constrained to lie in the epipolar plane defined by the ray in the first view $\mathbf{p}$ and the viewer translation $\mathbf{U}$ (figure 4.4):

$$\mathbf{p}_t = \frac{(\mathbf{U} \wedge \mathbf{p}) \wedge \mathbf{p}}{\lambda}. \tag{4.27}$$

In terms of measurements on the image sphere:

$$\mathbf{q}_t = \frac{(\mathbf{U} \wedge \mathbf{q}) \wedge \mathbf{q}}{\lambda} - \Omega \wedge \mathbf{q}, \tag{4.28}$$

where $\mathbf{q}_t$ is the image velocity of a point on the space curve at a distance $\lambda$. Equation (4.28) is the well known equation of structure from motion [149]. Points on successive image curves are "matched" by searching along epipolar great circles on the image sphere (or epipolar lines for planar image geometry) defined by the viewer motion, $\mathbf{U}$, $\Omega$ and the image position $\mathbf{q}$. Note also that the image velocity consists of two components. One component is determined purely by the viewer's rotational velocity about camera centre and is independent of the structure of the scene ($\lambda$). The other component is determined by the translational velocity of the viewer.

## 4.3.1    Depth from image velocities

As with apparent contours, depth $\lambda$ (distance along the ray $\mathbf{p}$) can be computed from the deformation ($\mathbf{p}_t$) of the image contour under known viewer motion. From (4.27):

$$\lambda = -\frac{\mathbf{U}.\mathbf{n}^p}{\mathbf{p}_t.\mathbf{n}^p}. \tag{4.29}$$

This formula is an infinitesimal analogue of triangulation with stereo cameras. The numerator is analogous to baseline and the denominator to disparity.

Equation (4.29) can also be re-expressed in terms of spherical image position $\mathbf{q}$ and the normal component of image velocity $\mathbf{q}_t.\mathbf{n}^p$:

$$\lambda = -\frac{\mathbf{U}.\mathbf{n}^p}{\mathbf{q}_t.\mathbf{n}^p + (\mathbf{\Omega} \wedge \mathbf{q}).\mathbf{n}^p}. \tag{4.30}$$

These equations are equivalent to those derived for an apparent contour. It is impossible to discriminate an apparent contour from a surface curve from instantaneous image velocities alone.

## 4.3.2 Curve tangent from rate of change of orientation of image tangent

Having recovered the depth of each point on the space curve it is possible to recover the geometry of the space curve by numerical differentiation. Here, an alternative method is presented. This recovers the curve tangent and normal directly from image measurables without first explicitly recovering the space curve. The space curve tangent, $\mathbf{T}$ can be recovered from the temporal derivative of the image curve tangent $\mathbf{t}^p$ as follows.

**Derivation 4.3** *Rearranging (4.11) we see that the space curve tangent only has components parallel to the ray and image tangent. Namely*

$$\mathbf{T} = [1 - (\mathbf{p}.\mathbf{T})^2]^{1/2}\mathbf{t}^p + (\mathbf{p}.\mathbf{T})\mathbf{p}. \tag{4.31}$$

*By differentiating 4.31 and looking at the normal components it is straightforward to show that the following relationship exists between the derivative of the image tangent and the space curve tangent:*

$$\mathbf{t}_t^p.\mathbf{n}^p = -\frac{(\mathbf{p}.\mathbf{T})}{[1 - (\mathbf{p}.\mathbf{T})^2]^{1/2}}\mathbf{p}_t.\mathbf{n}^p. \tag{4.32}$$

*This equation can be used to recover the coefficients of $\mathbf{t}^p$ and $\mathbf{p}$ in (4.31) and hence allows the recovery of the curve tangent. Alternatively it is easy to see from (4.31) that the following simple condition must hold:*

$$\mathbf{T}.\mathbf{n}^p = 0. \tag{4.33}$$

*Differentiating (4.33) with respect to time $t$ gives*

$$\mathbf{T}.\mathbf{n}_t^p = 0 \tag{4.34}$$

*since the space curve is assumed static and does not change with time. Equations (4.33) and (4.34) allow the recovery of the space curve tangent up to an arbitrary sign. In terms of measurements on the image sphere:*

$$\mathbf{T} = \frac{\tilde{\mathbf{n}}^p \wedge (\tilde{\mathbf{n}}_t^p + \Omega \wedge \tilde{\mathbf{n}}^p)}{|\tilde{\mathbf{n}}^p \wedge (\tilde{\mathbf{n}}_t^p + \Omega \wedge \tilde{\mathbf{n}}^p)|}. \tag{4.35}$$

The orientation of the space curve tangent is recovered from the change in the image curve normal and knowledge of the viewer's rotational velocity. A similar expression to (4.35) was derived by Faugeras et al. [161] for the image motion of straight lines.

### 4.3.3   Curvature and curve normal

We now show how to recover the space curve's curvature, $\kappa$ and normal, $\mathbf{N}$, directly from measurements on the spatio-temporal image and known viewer motion.

To simplify the derivation we choose a frame aligned with the curve normal with basis vectors $\{\mathbf{n}^p, \mathbf{n}_t^p, \mathbf{T}\}$. (It is easy to see from (4.33) and (4.34) that these three vectors are orthogonal. $\mathbf{n}_t^p$ is not necessarily a unit vector.) In this frame $\kappa\mathbf{N}$ can be expressed as:

$$\kappa\mathbf{N} = \alpha\mathbf{n}^p + \beta\mathbf{n}_t^p + \gamma\mathbf{T}. \tag{4.36}$$

From the definition of a space curve normal, $\gamma$ must be zero. The other two orthogonal components of $\kappa\mathbf{N}$, $\kappa\mathbf{N}.\mathbf{n}^p$ and $\kappa\mathbf{N}.\mathbf{n}_t^p$, can be recovered from the curvature in the image (4.20) and its temporal derivative as follows.

**Derivation 4.4** *By rearranging (4.20) we can solve for $\alpha$:*

$$\kappa^p = \lambda\kappa\frac{\mathbf{N}.\mathbf{n}^p}{(1 - (\mathbf{p}.\mathbf{T})^2)}$$

$$\alpha = \frac{\kappa^p}{\lambda}(1 - (\mathbf{p}.\mathbf{T})^2). \tag{4.37}$$

*Differentiating (4.20) and rearranging we can recover the other component, $\beta$:*

$$\kappa_t^p = -2\kappa^p\frac{\mathbf{U}.\mathbf{t}^p}{\lambda}\left[\frac{(\mathbf{p}.\mathbf{T})}{(1 - (\mathbf{p}.\mathbf{T})^2)^{1/2}}\right] - \kappa^p\frac{\mathbf{U}.\mathbf{p}}{\lambda} + \lambda\kappa\frac{\mathbf{N}.\mathbf{n}_t^p}{(1 - (\mathbf{p}.\mathbf{T})^2)} \tag{4.38}$$

$$\beta = \left[\frac{(1 - (\mathbf{p}.\mathbf{T})^2)}{\lambda}\right]\left[\kappa_t^p + +\kappa^p\frac{\mathbf{U}.\mathbf{p}}{\lambda}\right] + 2\kappa^p\frac{\mathbf{U}.\mathbf{t}^p}{\lambda^2}(\mathbf{p}.\mathbf{T})(1 - (\mathbf{p}.\mathbf{T})^2) \tag{4.39}$$

The space curve normal and curvature can be recovered directly from measurements in the image and known viewer motion.

For the analysis of the next section a simpler expression for the temporal derivative of the image curve's curvature is introduced. This is obtained by differentiating (4.17) and substituting (4.27) and (4.17):

$$\kappa_t^p = -\frac{\kappa \mathbf{B}.\mathbf{U}}{(1-(\mathbf{p}.\mathbf{T})^2)^{3/2}} - \frac{3\kappa^p\left[(\mathbf{p}.\mathbf{T})(\mathbf{U}.\mathbf{t}^p)\right]}{\lambda(1-(\mathbf{p}.\mathbf{T})^2)^{1/2}}. \tag{4.40}$$

In the special case of viewing a section of curve which projects to an inflection, i.e. $\kappa^p = 0$, or of the viewer moving in a direction which is perpendicular to the image curve's tangent, i.e. $\mathbf{U}.\mathbf{t}^p = 0$, the second term is zero and

$$\kappa_t^p = -\frac{\kappa \mathbf{B}.\mathbf{U}}{(1-(\mathbf{p}.\mathbf{T})^2)^{3/2}} \tag{4.41}$$

i.e. the sign of the deformation of the image curve encodes the sign of $\mathbf{B}.\mathbf{U}$ (since $\kappa$ and the denominator are always positive). This is sufficient to determine what the curve normal is doing qualitatively, i.e. whether the curve is bending towards or away from the viewer. This information is used in section 4.4.3 to recover qualitative properties of the underlying surface's shape.

If the image curve at time $t$ has a zero of curvature at $\mathbf{p}(s_0, t)$ because $\mathbf{p}$ lies in the osculating plane, the inflection at $s_0$ will not disappear in general under viewer motion but will move along the curve (see later, figure 4.9). Generically inflections can only be created or annihilated in pairs [42].

## 4.4 Surface geometry

### 4.4.1 Visibility constraint

Since the curve is visible the angle between the surface normal and the line of sight must be less than or equal to $90°$ (otherwise the local tangent plane would occlude the curve). If the angle is $90°$ then the image curve is coincident with the apparent contour of the surface.

The visibility constraint can be utilised to constrain surface orientation. Since,

$$-1 \leq \mathbf{p}.\mathbf{n} \leq 0, \tag{4.42}$$

if $\mathbf{p}$ is taken as the south pole of the Gaussian sphere, then the surface normal must lie on the northern hemisphere [4]. Each position of the viewer generates a fresh constraint hemisphere (see figure 4.5a). For known viewer movements these hemispheres can be intersected and the resultant patch on the Gaussian

---

[4] The convention used is that the surface normal is defined as being outwards from the solid surface.

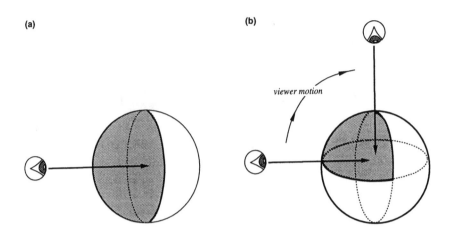

Figure 4.5: The visibility constraint.

*(a) The visibility constraint restricts the surface normal **n** to lie on a hemisphere of the (Gauss map) Gaussian sphere. (b) By intersecting these constraint regions for known viewer motions a tighter constraint is placed on the normal.*

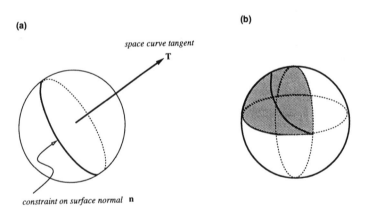

Figure 4.6: The tangency constraint.

*(a) The tangent constraint restricts the surface normal to lie on a great circle of the Gaussian sphere. (b) By intersecting this curve with the constraint patch from the visibility constraint, the surface normal is further restricted to an arc of a great circle.*

sphere places a tighter constraint on $\mathbf{n}$ (figure 4.5b). Clearly, motion which rotates by $180^{\circ}$ about the object determines the surface normal, provided, of course, the point is always visible. The point will not be visible if another part of the surface occludes it, or if it has reached the extremal boundary (the back projection of the apparent contour in the image). However, when it reaches the extremal boundary the normal is fully determined [17].

The viewer motion must be accurately known in order to fully utilise the visibility constraint over a sequence of views. Uncertainty in motion could be included in a primitive fashion by intersecting regions larger than a hemisphere. The excess over a hemisphere would be bounded by estimates of error in viewer motion.

The constraint on the normal provided by the visibility constraint is applicable to texture points as well as smooth curves. The following constraint exploits the continuity of the curve.

### 4.4.2 Tangency constraint

The space curve tangent lies in the surface tangent plane and this constrains the surface normal $\mathbf{n}$:

$$\mathbf{T}.\mathbf{n} = 0. \tag{4.43}$$

This orthogonality condition generates a constraint curve which is a great circle on the Gaussian sphere (figure 4.6a) [23]. If the curve tangent could be determined exactly, then intersecting the great circle with the constraint patch from the visibility constraint would restrict the normal to an arc of a great circle (figure 4.6b). In practice there will be errors in the tangent so the constraint region will be a band rather than a curve. Combining information from many views will more accurately determine the tangent (and hence the constraint band). However, no "new" information is generated in each view as it is using the visibility constraint.

### 4.4.3 Sign of normal curvature at inflections

Even if the curvature and Frenet frame of a space curve lying on a surface are known, no constraint is placed on the surface curvature because the relation of the surface normal to the curve's osculating plane is unknown and arbitrary. However, it is shown below that at an inflection in the image curve, the sign of the normal curvature along the curve can be determined without first determining the surface normal. Moreover it can be determined without having to recover the space curve. It is shown that by following the inflection through a sequence of images, the sign of the normal curvature is determined along the curve. This can be done with incomplete, qualitative knowledge of viewer motion. The

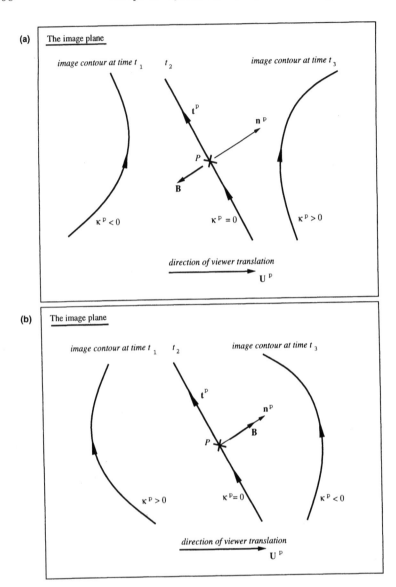

Figure 4.7: Orientation of the space curve from the deformation at inflections.

*If the viewer crosses the osculating plane of the surface curve the curvature of the image will change sign – projecting to an inflection when the viewer lies in osculating plane (time $t_2$). Two possible cases are shown.*

*(a) If $\mathbf{N} \cdot \mathbf{p} > 0$ the space curve is bending away from the viewer (figure 7a) and the image curve changes locally from an arc with negative image curvature via an inflection to an arc with positive image curvature, i.e. $\kappa_t^p > 0$.*

*(b) If $\mathbf{N} \cdot \mathbf{p} < 0$ the space curve is bending towards the viewer (figure 4.8b) and the opposite transition is seen in the image, i.e. $\kappa_t^p < 0$.*

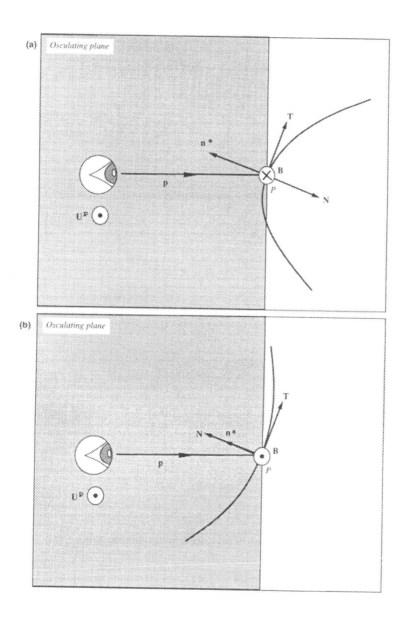

Figure 4.8: Sign of normal curvature from osculating plane geometry

*The component of the surface normal in the osculating plane ($\mathbf{n}^\star$) is constrained to lie within 90° of the ray, $\mathbf{p}$ by the visibility constraint ($\mathbf{n}.\mathbf{p} \leq 0$, shown shaded) and must be perpendicular to the surface curve tangent, $\mathbf{T}$ by the tangency constraint ($\mathbf{T} \cdot \mathbf{n} = 0$). The component of the surface normal in the osculating plane will therefore be parallel to the curve normal $\mathbf{N}$ – either in opposite directions (a) if $\mathbf{N} \cdot \mathbf{p} > 0$, or (b) in the same direction if $\mathbf{N} \cdot \mathbf{p} < 0$. The sign of the normal curvature in the direction of the curve tangent is determined by the sign of $\mathbf{N} \cdot \mathbf{p}$ which is obtained by noting the transition in image curvature at an inflection (figure 4.7).*

only requirement is to know whether the viewer is translating to the left or right of the image contour. This result is a generalisation to surface curves of Weinshall's [212] result for surface texture.

**Theorem 4.1 (Sign of normal curvatures at inflections)**   *Consider two views of a curve lying on a smooth surface. If a point P on the curve projects to an inflection in one view, but not in the other, then the normal curvature (in the direction of the curve's tangent) at P is convex [5] (concave) if the image of the curve at P has positive (negative) image curvature, $\kappa^p$, in the second view. The parameterisation of the curve is chosen so that $\mathbf{U}^p \cdot \mathbf{n}^p > 0$, where $\mathbf{U}^p$ is the projection in the image of the translational velocity of the viewer. With this parameterisation $\mathrm{sign}(\kappa^n) = -\mathrm{sign}(\kappa_t^p)$ where $\kappa^n$ is the normal curvature and $\kappa_t^p$ the time derivative of image curvature at the inflection.*

The proof below is in three stages. First, the viewing geometry is established (curve viewed in osculating plane, so it may be thought of as a plane curve). This determines $\mathbf{B} = \pm\mathbf{n}^p$ and hence constrains $\mathbf{N}$. Second, the sign of $\mathbf{N} \cdot \mathbf{p}$ is determined from the time derivative of image curvature (this determines whether the curve bends towards or away from the viewing direction), see figure 4.7. Third, the visibility constraint is utilised to relate space curve curvature (the bending) to surface normal curvature, see figure 4.8.

**Parameterisation**   The directions of the tangent vectors $\mathbf{T}$, $\mathbf{t}^p$ and the image curve normal, $\mathbf{n}^p$, are arbitrarily defined by choosing the parameterisation direction for the image curve. A convenient parameterisation is to choose the direction of the curve normal so that it is on the same side of the image contour as the projection of the translational velocity, i.e. $\mathbf{U}^p \cdot \mathbf{n}^p > 0$. This is always a good choice since it only fails when $\mathbf{U}^p \cdot \mathbf{n}^p = 0$ in which case the viewer is moving in the osculating plane and both views contain an inflection for the same point on the space curve. Since $\{\mathbf{p}, \mathbf{t}^p, \mathbf{n}^p\}$ form an orthonormal right-handed system (with $\mathbf{p}$ into the image plane), fixing the direction of the curve normal $\mathbf{n}^p$ also fixes $\mathbf{t}^p$ and hence $\mathbf{T}$ and the sign of $\kappa^p$.

**Proof**

We first establish a relation between $\mathbf{N} \cdot \mathbf{p}$ and $\mathbf{B} \cdot \mathbf{n}^p$. From (4.11):

$$\mathbf{T} = \alpha \mathbf{t}^p + \beta \mathbf{p}, \quad \text{with} \quad \alpha > 0 . \tag{4.44}$$

---

[5] If we define the surface normal as being outwards from the solid surface, the *normal* curvature will be negative in any direction for a convex surface patch.

Then

$$\mathbf{N} \cdot \mathbf{p} = \mathbf{p} \cdot (\mathbf{B} \wedge \mathbf{T})$$
$$= \mathbf{B} \cdot (\mathbf{T} \wedge \mathbf{p})$$
$$= \alpha \mathbf{B} \cdot (\mathbf{t}^p \wedge \mathbf{p})$$
$$= -\alpha \mathbf{B} \cdot \mathbf{n}^p \qquad (4.45)$$

The last steps following from (4.44) and (4.15), and since $\alpha > 0$

$$\text{sign}(\mathbf{N} \cdot \mathbf{p}) = -\text{sign}(\mathbf{B} \cdot \mathbf{n}^p) \qquad (4.46)$$

1. **Osculating plane constraints**

   If a point on a surface curve projects to an inflection in one view, but not in another then (from section 4.2.3) the ray in the first view must lie in the osculating plane and from (4.17)

$$\mathbf{B} \cdot \mathbf{p} = 0 \qquad (4.47)$$

   Since $\mathbf{B} \cdot \mathbf{T} = 0$ we have from (4.44)

$$\mathbf{B} \cdot \mathbf{t}^p = 0 \qquad (4.48)$$

   Thus, using the orthogonal triad $\{\mathbf{p}, \mathbf{t}^p, \mathbf{n}^p\}$

$$\mathbf{B} = \pm(\mathbf{p} \wedge \mathbf{t}^p) \qquad (4.49)$$
$$= \pm \mathbf{n}^p \qquad (4.50)$$

2. **Sign of $\mathbf{N} \cdot \mathbf{p}$**

   The transition in image curvature at $P$ from a point of inflection in the first view ($\kappa^p = 0$) to an arc with positive or negative image curvature, $\kappa^p$, in the second view determines the sign of $\mathbf{N} \cdot \mathbf{p}$ (figure 4.7).

   We can express $\mathbf{U}$ in the orthogonal triad: $\{\mathbf{p}, \mathbf{t}^p, \mathbf{n}^p\}$

$$\mathbf{U} = \gamma \mathbf{n}^p + \delta \mathbf{t}^p + \epsilon \mathbf{p}, \quad \text{with} \quad \gamma > 0; \qquad (4.51)$$

   the sign of $\gamma$ follows from the parameterisation choice that $\mathbf{U}^p \cdot \mathbf{n}^p > 0$. Using (4.47) and (4.48) gives:

$$\mathbf{B} \cdot \mathbf{U} = \gamma \mathbf{B} \cdot \mathbf{n}^p \qquad (4.52)$$

   and hence from (4.41) (noting that $\kappa$ and the denominator are positive):

$$\text{sign}(\kappa_t^p) = -\text{sign}(\mathbf{B} \cdot \mathbf{U})$$
$$= -\text{sign}(\mathbf{B} \cdot \mathbf{n}^p)$$
$$= \text{sign}(\mathbf{N} \cdot \mathbf{p}) \qquad (4.53)$$

   the last step following from (4.46). This result determines the orientation of the curve normal, $\mathbf{N}$, relative to the line of sight (figure 4.8).

3. **Sign of** $\kappa^n$

We express the surface normal $\mathbf{n}$ in the orthogonal triad $\{\mathbf{T}, \mathbf{N}, \mathbf{B}\}$:

$$\mathbf{n} = \hat{\alpha}\mathbf{N} + \hat{\beta}\mathbf{B} \qquad (4.54)$$

since from (4.43) $\mathbf{T} \cdot \mathbf{n} = 0$. Hence, $\mathbf{p} \cdot \mathbf{n} = \hat{\alpha}\mathbf{N} \cdot \mathbf{p}$ since from (4.47) $\mathbf{B} \cdot \mathbf{p} = 0$. The visibility constraint restricts the sign as $\mathbf{p} \cdot \mathbf{n} \leq 0$, and hence $\text{sign}(\mathbf{N} \cdot \mathbf{p}) = -\text{sign}(\hat{\alpha})$. The sign of the normal curvature $\kappa^n$ then follows from the above and (4.53):

$$
\begin{aligned}
\text{sign}(\kappa^n) &= \text{sign}(\mathbf{N} \cdot \mathbf{n}) \\
&= \text{sign}(\hat{\alpha}) \\
&= -\text{sign}(\mathbf{N} \cdot \mathbf{p}) \\
&= -\text{sign}(\kappa_t^p)\square \qquad (4.55)
\end{aligned}
$$

Note that the test is only valid if the inflection in the first view moves along the image curve in the next since an inflection corresponding to the same point on the surface curve in both views can result from either zero normal curvature or motion in the osculating plane. Two views of a surface curve are then sufficient to determine the sign of the normal curvature. However, because of the numerical difficulties in determining a zero of curvature, the test can be applied with greater confidence if a transition from (say) negative to zero (an inflection) to positive image curvature is observed. The component of viewer translation parallel to the image plane is only used to determine the direction of the curve parameterisation. No knowledge of viewer rotations are required. The theorem is robust in that only partial knowledge (or inaccurate knowledge but with bounded errors) of translational velocity will suffice. This can be estimated from image measurements by motion parallax [138, 182] or is readily available in the case of binocular vision (where the camera or eye positions are constrained).

## Applications

Figures 4.9 – 4.11 show examples of the application of this result to real images.

1. **Determining the sign of normal curvature by tracking inflections**

   Figure 4.9 shows a sequence of images taken by a CCD camera mounted on an Adept robot arm rotating around a vase (figure 4.1). Two image curves are selected – the projection of surface curves on the neck (hyperbolic patch) and body (elliptic patch) of the vase respectively. These image curves are automatically tracked using B-spline snakes [50]. Crosses mark inflection points. As the viewer moves from left to right the inflections

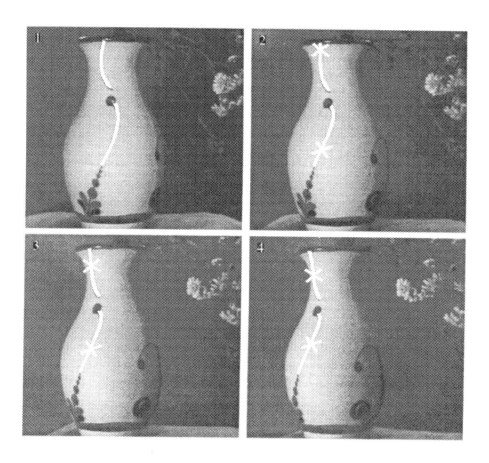

Figure 4.9: Tracking inflections to determine the sign of normal curvature.
*Four images are shown from an image sequence taken by a camera moving (from left to right with fixation) around a smooth surface (a vase). The image contours are tracked by using B-spline snakes. Inflections (marked by a cross) are generated for points whose osculating plane contains the vantage point. Under viewer motion the preimage of the inflection moves along the surface curve. The change in the sign image curvature is sufficient to determine the sign of the normal curvature along the curve. For the top part of the curve on the neck of the vase the transition in image curvature is from positive to negative indicating concave normal sections along the curve. For the bottom part of the curve on the body of the vase the transition is the opposite indicating convex normal sections. This information is consistent with the neck of the vase being hyperbolic and the body convex elliptic. Note that this classification has been achieved with partial knowledge of the viewer motion. The only knowledge required is whether the component of viewer translation parallel to the image plane is to the left or right of the image contours.*

Figure 4.10: Qualitative information for grasping.

*The left and right images of a Japanese tea cup are shown.  The simple test can be used for the Chinese characters painted in the indentations created by the potter's thumb imprints.  The transition in the sign of image curvature indicates a concave section.  These thumb imprints are created to aid in grasping the tea cup.*

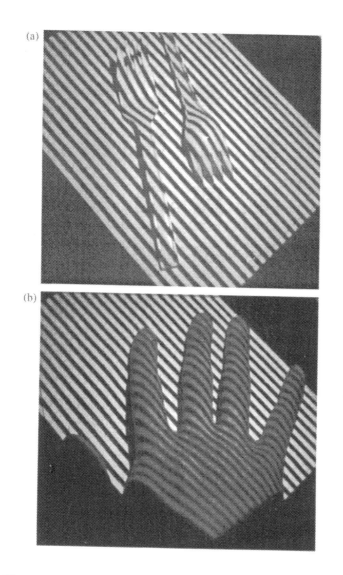

Figure 4.11: Qualitative interpretation of rangefinder images.

*These images were formed by the projection of planes of light onto the workspace and viewing the scene with a camera whose centre is displaced to the left of the light plane. The surface is covered artificially with a set of planar curves. If these surface curves were viewed in their respective osculating plane (the light planes) their images would simply be straight lines (degenerate case of an inflection). The sign of the image curvature in the images shown is consistent with the sign of the normal curvatures along the curve. Positive, negative and zero image curvature indicate respectively convex, concave and flat sections.*

moves smoothly along each curve. For the curve on the hyperbolic section of the vase the transition in image curvature is from positive to negative indicating a concave normal section. For the curve on the elliptic section the transition is the reverse indicating a convex normal section. The main advantage is that provided the position of the inflection in one view can be identified with a convex or concave arc in the second – say by a nearby texture point – then only partial knowledge of the viewer motion is needed. In the case shown the only knowledge required is whether the viewer translation is to the left or right of the image contours.

2. **Convexity/Concavity test of surface**
The test directly rules out certain types of surfaces as follows. If the normal curvature is negative then the surface could be convex or hyperbolic – it cannot be concave. Similarly, a positive normal curvature rules out a convex surface. Thus the result can be used as a test for non-convexity or non-concavity. This is similar to the information available from the motion of a specularity when the light source position is not known [220].

3. **Combination with other cues**
The test is most powerful when combined with other cues. Extremal boundaries, for example, are an extremely rich source of surface shape information. Unfortunately they cannot provide information on concave surface patches since these will never appear as extremal boundaries. The information available from the deformation of other curves is therefore extremely important even though it is not as powerful a cue as the image of the extremal boundary. For example at an extremal boundary the Gaussian curvature of the surface is known from the sign of the curvature of the apparent contour in the image. Consider an elliptic region (this must be convex to appear on the extremal contour), if there is a point $P$ inside the extremal boundary with concave (positive) normal curvature (determined using the test above), then there must be at least one parabolic curve between $P$ and the boundary.

This information can be used, for example, to indicate the presence of concavities for grasping. Figure 4.10 shows an example of a Japanese tea cup. The application of the test to the image contours of the Chinese characters in the two indentations indicate the presence of concavities. These concavities (thumb imprints) are deliberately placed by Japanese potters when making tea cups since they are good for grasping the cup.

4. **Interpreting single images of planar curves**
It is well known that by making certain assumptions about the nature of

the surface curves humans can interpret the shape of visible surfaces from single images of surface curves. Typically this has been used to infer the orientation of planar surfaces. Stevens [188] has shown how the assumptions of parallel contour generators (with the implicit assumption that the curvature perpendicular to the contour is a principal curvature with zero curvature) can be used to recover the shape of curved surfaces. The result described above can be used to make precise the intuition of Stevens that the appearance of planar surface curves is sufficient to tell us a lot about the shape of the surface. This is highlighted with a simple example based on the qualitative interpretation of range finder images. Figure 4.11 shows two images taken with a rangefinder system [29]. The images are formed by the projection of planes of light on to the workspace and viewing this scene with another camera whose centre is displaced away from the light plane. The effect is to cover the untextured surface with planar surface curves. The osculating plane of these curves is simply the light plane. If the camera is placed in the osculating plane the light stripes would appear as straight lines. (The straight line is simply a degenerate case of an inflection.) By taking an image with the camera on one side of the osculating plane (the light plane) the straight lines deform in a way determined exactly as predicted by the test above. The sign of the image curvature is consistently related to the sign of the normal curvature along the curve. In the examples of figure 4.11 the camera is displaced to the right of the light plane projector and so convexities (negative image curvature), concavities (positive image curvature) and inflections (zero image curvature) indicate respectively convex, concave and zero normal curvatures. The cue is extremely powerful, giving a strong sense of the shape of the visible surfaces without the need for accurate image measurements, exact epipolar geometry and triangulation which are required in the established, quantitative approach for interpreting rangefinder images.

### 4.4.4 Surface curvature at curve intersections

If surface curves cross transversely, or the curve's tangent is discontinuous, more constraints can be placed on the surface geometry. In principle from two views it is possible to reconstruct both space curves and hence determine their tangents and curvatures at the intersection. (Equations (4.29), (4.31), (4.37) and (4.39) show how these can be estimated directly from $\{\mathbf{p}, \mathbf{t}^p, \mathbf{n}^p\}$ and their temporal derivatives under known viewer motion.) From these the normal curvatures $\kappa^{n(1)}, \kappa^{n(2)}$ along the two tangent directions can be determined

$$\kappa^{n(1)} = \kappa^{(1)} \mathbf{N}^{(1)} . \mathbf{n} \tag{4.56}$$

$$\kappa^{n(2)} \quad = \quad \kappa^{(2)} \mathbf{N}^{(2)} . \mathbf{n} \tag{4.57}$$

where $\mathbf{n}$ is the surface normal

$$\mathbf{n} = \frac{\mathbf{T}^{(1)} \wedge \mathbf{T}^{(2)}}{\|\mathbf{T}^{(1)} \wedge \mathbf{T}^{(2)}\|} \tag{4.58}$$

and $\kappa^{(i)}, \mathbf{T}^{(i)}, \mathbf{N}^{(i)}$  $i = 1, 2$ are the curvature, tangent and normals of the space curves at the intersection. An intersection of the two curves ensures sufficient information to recover the surface normal. It has the added advantage that the exact epipolar geometry is not required to match points in the two images since the intersection can be tracked. It also has the advantage that the surface normal can be computed by measuring the change in image curve normals and knowledge of camera rotation alone. This is done by applying (4.35) to recover both space curve tangents and taking their vector product to determine the surface normal (4.58). However, the recovery of two normal curvatures is not sufficient in general to determine the Gaussian curvature of the surface. (There are many convex and concave directions on a hyperbolic patch.) It can only constrain its sign. The problem is that although the angle between the tangent vectors is known, the relation between the tangent pair and the principle directions is unknown. From Euler's formula [67] we have

$$\kappa^{n(1)} \quad = \quad \kappa_1 \sin^2 \theta + \kappa_2 \cos^2 \theta \tag{4.59}$$
$$\kappa^{n(2)} \quad = \quad \kappa_1 \sin^2 (\theta + \alpha) + \kappa_2 \cos^2 (\theta + \alpha) \tag{4.60}$$

where $\alpha$ is the angle between the tangents in the tangent plane; $\theta$ is the (unknown) angle between the tangent, $\mathbf{T}^{(1)}$, and *principal* direction; $\kappa_1$ and $\kappa_2$ are the principal curvatures. There are three unknowns, $\kappa_1, \kappa_2, \theta$, and only two constraints. If there is a triple crossing or higher there are sufficient constraints to uniquely determine the three unknowns. This is less likely to occur. However, we can catalogue the surface by the sign of the Gaussian curvature.

| Sign of $\kappa^{n(1)}$ and $\kappa^{n(2)}$ | Surface |
|---|---|
| both negative | *not* concave |
| both positive | *not* convex |
| one negative, one positive | hyperbolic |

Furthermore, it is easy to show that there is a lower bound on the difference of the principle curvatures, namely

$$|\kappa_1 - \kappa_2| \geq \left| \frac{\kappa^{n(1)} - \kappa^{n(2)}}{\sin \alpha} \right|, \tag{4.61}$$

where $\alpha$ is the angle between the tangents in the tangent plane and $\kappa^{n(i)}$ are measured normal curvatures.

**Derivation 4.5** *Subtracting the two copies of (4.59) for the curvatures of the two curves and from trigonometry:*

$$\kappa^{n(1)} - \kappa^{n(2)} = (\kappa_1 - \kappa_2)(\sin \alpha \sin (\alpha + 2\theta)). \qquad (4.62)$$

*Rearranging and inspecting the magnitude of both sides gives the required result.*

This is simply repackaging the information contained in the normal curvatures and the angle between tangents. However, in this form it is better suited to a first filter on a model data base. For example if it were known that the model was convex a negative sign test would reject it. Similarly if the principle curvatures were known not to exceed certain limits, then the lower bound test might exclude it.

## 4.5 Ego-motion from the image motion of curves

We have seen that under the assumption of known viewer motion constraints on surface shape can be derived from the deformation of image curves. It is now shown that despite the aperture problem the image motion of curves can provide constraints on ego-motion.

Equation (4.28) relates the image velocity of a point on a curve in space to the viewer's motion. In principle measurement of image velocities at a finite number of discrete points can be used to solve for the viewer's ego-motion [175, 138, 159, 149, 150] – the five unknowns include the three components of rotational velocity $\mathbf{\Omega}$ and the direction of translation $\mathbf{U}/|\mathbf{U}|$. Once the ego-motion has been recovered it is then possible to recover the structure (depth of each point) of the scene. This is the well known structure from motion problem. [6]

The structure from motion problem relies on the ability to measure the image velocities of a number of points. Although this is possible for distinct, discrete points – "corner features" – this is impossible for a point on an image curve from purely local measurements. Measuring the (real) image velocity $\mathbf{q}_t$ for a point on an image curve requires knowledge of the viewer motion – equation (4.28). Only the normal component of image velocity (vernier velocity) can be obtained directly from local measurements at a curve – the *aperture problem* [201].

---

[6] In practice this is an extremely difficult and ill-conditioned problem. Apart from the obvious *speed–scale* ambiguity which makes it impossible to recover the magnitude of the translational velocity (and hence to recover absolute depth) there is a problem with the *bas–relief* ambiguity. This is the problem of differentiating between a deeply indented surface rotating through a small angle, and a shallowly indented surface rotating through a larger angle. That is, not all of the variables to be estimated will be well conditioned. Despite these drawbacks some attempts at the problem of structure from motion problem have been extremely successful. An excellent example is the system developed by Harris and colleagues [93, 96, 49] which tracks image corner features over time to solve for the ego-motion and to reconstruct the scene.

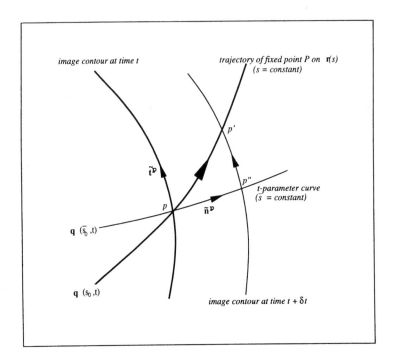

Figure 4.12: Spatio-temporal parameterisations of the image.

*p is the projection of a point P ($\mathbf{r}(s_0)$) on a space curve, $\mathbf{r}(s)$. Under viewer motion the image of P follows a trajectory $\mathbf{q}(s_0, t)$ in the spatio-temporal image. The tangent to this trajectory is the instantaneous image velocity, $\mathbf{q}_t$ – shown as a vector pp'. The image velocity can be decomposed into two orthogonal components: the normal component perpendicular to the image contour (vector pp'') and the tangential component parallel to the image tangent (vector p''p'). From local image measurements and without knowledge of viewer motion it is impossible to measure the real image velocity. It is only possible to measure the normal component. An alternative parameterisation is to define the t–parameter curves to be orthogonal to the image contours. Such a parameterisation can always be set up in the spatio-temporal image and it is independent of viewer motion.*

Knowledge of the normal component of image velocity alone is insufficient to solve for the ego-motion of the viewer. By assuming no (or knowledge of) rotational velocity qualitative constraints can be recovered [106, 186]. By making certain assumptions about the surface being viewed a solution may sometimes be possible. Murray and Buxton [158] show, for example, how to recover ego-motion and structure from a minimum of eight vernier velocities from the same planar patch.

In the following we show that it is also possible to recover ego-motion without segmentation or making any assumption about surface shape. The only assumption made is that of a static scene. The only information used is derived from the spatio-temporal image of an image curve under viewer motion. This is achieved by deriving an additional constraint from image accelerations. This approach was motivated by the work of Faugeras [71] which investigated the relationship between optical flow and the geometry of the spatio-temporal image. In the following analysis a similar result is derived independently. Unlike Faugeras's approach the techniques of differential geometry are not applied to the spatio-temporal image surface. Instead the result is derived directly from the equations of the image velocity and acceleration of a point on a curve by expressing these in terms of quantities which can be measured from the spatio-temporal image. The derivation follows.

The image velocity of a point on a fixed space curve is related to the viewer motion and depth of the point by (4.28):

$$q_t = \frac{(\mathbf{U} \wedge \mathbf{q}) \wedge \mathbf{q}}{\lambda} - \mathbf{\Omega} \wedge \mathbf{q}.$$

By differentiating with respect to time and substituting the *rigidity* constraint [7]

$$\lambda_t + \mathbf{q}.\mathbf{U} = 0 \tag{4.63}$$

the normal component of acceleration can be expressed in terms of the viewer's motion, $(\mathbf{U}, \mathbf{U}_t, \mathbf{\Omega}, \mathbf{\Omega}_t)$, and the 3D geometry of the space curve $(\lambda)$ [8],

$$q_{tt}.\tilde{\mathbf{n}}^p = -\frac{\mathbf{U}_t.\tilde{\mathbf{n}}^p}{\lambda} + \frac{(\mathbf{U}.\mathbf{q})(q_t.\tilde{\mathbf{n}}^p)}{\lambda} - \frac{(\mathbf{q}.\mathbf{U})(\mathbf{U}.\tilde{\mathbf{n}}^p)}{(\lambda)^2}$$
$$+ (\mathbf{\Omega}_t.\tilde{\mathbf{t}}^p) + ((\mathbf{\Omega}.\mathbf{q}) \left[ \frac{\mathbf{U}.\tilde{\mathbf{t}}^p}{\lambda} + (\mathbf{\Omega}.\tilde{\mathbf{n}}^p) \right]. \tag{4.64}$$

Note that because of the aperture problem neither the image velocity $q_t$ nor the image acceleration $q_{tt}.\tilde{\mathbf{n}}^p$ can be measured directly from the spatio-temporal im-

---

[7] Obtained by differentiating (4.10) and using the condition (4.26) that for a fixed space curve, $\mathbf{r}_t = 0$.

[8] This is equivalent to equation (2.47) derived for the image acceleration at an apparent contour where because we are considering a rigid space-curve, $1/\kappa^t = 0$.

age. Only the normal component of the image velocity, $\mathbf{q}_t.\tilde{\mathbf{n}}^p$, (vernier velocity) can be directly measured.

Image velocities and accelerations are now expressed in terms of measurements on the spatio-temporal image. This is achieved by re-parameterising the image so that it is independent of knowledge of viewer motion. In the epipolar parameterisation of the spatio-temporal image, $\mathbf{q}(s,t)$, the $s$-parameter curves were defined to be the image contours while the $t$-parameter curves were defined by equation (4.28) to be the trajectory of the image of a fixed point on the space curve. At any instant the magnitude and direction of the tangent to a $t$-parameter curve is equal to the (real) image velocity, $\mathbf{q}_t$ – more precisely $\frac{\partial \mathbf{q}}{\partial t}\big|_s$. Note that this parameter curve is the trajectory in the spatio-temporal image of a fixed point on the space curve if such a point could be distinguished.

A parameterisation can be chosen which is completely independent of knowledge of viewer motion, $\mathbf{q}(\bar{s},t)$, where $\bar{s} = \bar{s}(s,t)$. Consider, for example, a parameterisation where the $t$-parameter curves (with tangent $\frac{\partial \mathbf{q}}{\partial t}\big|_{\bar{s}}$) are chosen to be orthogonal to the $\bar{s}$-parameter curves (with tangent $\frac{\partial \mathbf{q}}{\partial \bar{s}}\big|_t$) – the image contours. Equivalently the $t$-parameter curves are defined to be parallel to the curve normal $\tilde{\mathbf{n}}^p$,

$$\frac{\partial \mathbf{q}}{\partial t}\Big|_{\bar{s}} = \beta\tilde{\mathbf{n}}^p \qquad (4.65)$$

where $\beta$ is the magnitude of the normal component of the (real) image velocity. The advantage of such a parameterisation is that it can always, in principle, be set up in the image without any knowledge of viewer motion. [9] The (real) image velocities can now be expressed in terms of the new parameterisation (see figure 4.12).

$$\mathbf{q}_t = \frac{\partial \mathbf{q}}{\partial t}\Big|_s \qquad (4.66)$$

$$= \frac{\partial \bar{s}}{\partial t}\Big|_s \frac{\partial \mathbf{q}}{\partial \bar{s}}\Big|_t + \frac{\partial \mathbf{q}}{\partial t}\Big|_{\bar{s}}. \qquad (4.67)$$

Equation (4.67) is simply resolving the (real) image velocity $\mathbf{q}_t$ into a tangential component which depends on $\left(\frac{\partial \bar{s}}{\partial t}\big|_s\right)$ (and is not directly available from the spatio-temporal image) and the normal component of image velocity $\beta$ which can be be measured.

$$\mathbf{q}_t = \frac{\partial \bar{s}}{\partial t}\Big|_s \left|\frac{\partial \mathbf{q}}{\partial \bar{s}}\Big|_t\right| \tilde{\mathbf{t}}^p + \beta\tilde{\mathbf{n}}^p. \qquad (4.68)$$

---

[9]Faugeras [71] chooses a parameterisation which preserves image contour arc length. He calls the tangent to this curve the apparent image velocity and he conjectures that this is related to the image velocity computed by many techniques that aim to recover the image velocity field at closed contours [100]. The tangent to the $t$-parameter curve defined in our derivation has an exact physical interpretation. It is the (real) normal image velocity.

The (real) image acceleration can be similarly expressed in terms of the new parameterisation.

$$\mathbf{q}_{tt} = \left. \frac{\partial^2 \mathbf{q}}{\partial^2 t} \right|_s \tag{4.69}$$

$$= \left. \frac{\partial^2 \bar{s}}{\partial^2 t} \right|_s \left. \frac{\partial \mathbf{q}}{\partial \bar{s}} \right|_t + \left( \left. \frac{\partial \bar{s}}{\partial t} \right|_s \right)^2 \left. \frac{\partial^2 \mathbf{q}}{\partial^2 \bar{s}} \right|_t + 2 \left. \frac{\partial \bar{s}}{\partial t} \right|_s \left. \frac{\partial}{\partial \bar{s}} \left( \left. \frac{\partial \mathbf{q}}{\partial t} \right|_{\bar{s}} \right) \right|_t + \left. \frac{\partial^2 \mathbf{q}}{\partial^2 t} \right|_{\bar{s}}$$

$$\mathbf{q}_{tt}.\tilde{\mathbf{n}}^p = \left( \left. \frac{\partial \bar{s}}{\partial t} \right|_s \right)^2 \left. \frac{\partial^2 \mathbf{q}}{\partial^2 \bar{s}} \right|_t .\tilde{\mathbf{n}}^p + 2 \left. \frac{\partial \bar{s}}{\partial t} \right|_s \left. \frac{\partial}{\partial \bar{s}} \left( \left. \frac{\partial \mathbf{q}}{\partial t} \right|_{\bar{s}} \right) \right|_t .\tilde{\mathbf{n}}^p + \left. \frac{\partial^2 \mathbf{q}}{\partial^2 t} \right|_{\bar{s}} .\tilde{\mathbf{n}}^p \tag{4.70}$$

Apart from $\left( \left. \frac{\partial \bar{s}}{\partial t} \right|_s \right)$ which we have seen determines the magnitude of the tangential component of image curve velocity (and is not measurable) the other quantities in the right-hand side of the (4.70) are directly measurable from the spatio-temporal image. They are determined by the curvature of the image contour, $\kappa^p$; the variation of the normal component of image velocity along the contour, $\left. \frac{\partial \beta}{\partial \bar{s}} \right|_t$; and the variation of the normal component of image velocity perpendicular to the image contour respectively, $\left. \frac{\partial \beta}{\partial t} \right|_{\bar{s}}$.

In equation (4.64) the normal component of image acceleration is expressed in terms of the viewer's motion, $(\mathbf{U}, \mathbf{U}_t, \mathbf{\Omega}, \mathbf{\Omega}_t)$, and the 3D geometry of the space-curve. Substituting for $\lambda$,

$$\lambda = -\frac{\mathbf{U}.\tilde{\mathbf{n}}^p}{\mathbf{q}_t.\tilde{\mathbf{n}}^p + (\mathbf{\Omega}.\tilde{\mathbf{t}}^p)} \tag{4.71}$$

the right hand side of equation (4.64) can be expressed completely in terms of the unknown parameters of the viewer's ego-motion.

In equation (4.70) the normal component of image acceleration is expressed in terms of measurements on the spatio-temporal image and the unknown quantity $\left. \frac{\partial \bar{s}}{\partial t} \right|_s$ which determines the magnitude of the tangential velocity. This is not, however, an independent parameter since from (4.28),(4.30) and (4.67) it can be expressed in terms of viewer motion:

$$\left. \frac{\partial \bar{s}}{\partial t} \right|_s = \frac{\mathbf{q}_t. \left. \frac{\partial \mathbf{q}}{\partial \bar{s}} \right|_t}{\left| \left. \frac{\partial \mathbf{q}}{\partial \bar{s}} \right|_t \right|^2} \tag{4.72}$$

$$= \frac{1}{\left| \left. \frac{\partial \mathbf{q}}{\partial \bar{s}} \right|_t \right|} \left[ \frac{\mathbf{U}.\tilde{\mathbf{t}}^p}{\mathbf{U}.\tilde{\mathbf{n}}^p} (\mathbf{q}_t.\tilde{\mathbf{n}}^p + (\mathbf{\Omega}.\tilde{\mathbf{t}}^p) ) - (\mathbf{\Omega}.\tilde{\mathbf{n}}^p) \right]. \tag{4.73}$$

The right hand side of equation (4.70) can therefore also be expressed in terms of the unknown parameters of the viewer motion only. Combining equations (4.64) and (4.70) and substituting for $\left. \frac{\partial \bar{s}}{\partial t} \right|_s$ and $\lambda$ we can obtain a polynomial equation in terms of the unknown parameters of the viewer's motion $(\mathbf{U}, \mathbf{U}_t, \boldsymbol{\Omega}, \boldsymbol{\Omega}_t)$ with coefficients which are determined by measurements on the spatio-temporal image – $\{\mathbf{q}, \widetilde{\mathbf{t}}^p, \widetilde{\mathbf{n}}^p\}$, $\kappa^p$, $\beta$, $\left.\frac{\partial \beta}{\partial \bar{s}}\right|_t$ and $\left.\frac{\partial \beta}{\partial t}\right|_{\bar{s}}$. A similar equation can be written at each point on any image curve and if these equations can be solved it may be possible, in principle, to determine the viewer's ego-motion and the structure of the visible curves.

Recent experimental results by Arbogast [4] and Faugeras and Papadopoulo [74] validate this approach. Questions of the uniqueness and robustness of the solution remain to be investigated. These were our prime reasons for not attempting to implement the method presented. The result is included principally for its theoretical interest – representing a solution for the viewer ego-motion from the image motion of curves. In the Chapter 5 we see that instead of solving the structure from motion problem completely, reliable and useful information can be efficiently obtained from qualitative constraints.

## 4.6  Summary

In this chapter the information available from an image curve and its deformation under viewer motion has been investigated. It was shown how to recover the differential geometry of the space curve and described the constraints placed on the differential geometry of the surface. It was also shown how the deformation of image curves can be used, in principle, to recover the viewer's ego-motion.

Surprisingly – even with exact epipolar geometry and accurate image measurements – very little quantitative information about local surface shape is recoverable. This is in sharp contrast to the extremal boundaries of curved surfaces in which a single image can provide strong constraints on surface shape while a sequence of views allows the complete specification of the surface. However the apparent contours cannot directly indicate the presence of concavities. The image of surface curves is therefore an important cue.

The information available from image curves is better expressed in terms of incomplete, qualitative constraints on surface shape. It has been shown that visibility of the curve constrains surface orientation and moreover that this constraint improves with viewer motion. Furthermore, tracking image curve inflections determines the sign of normal curvature along the surface curve's tangent. This can also be used to interpret the images of planar curves on surfaces – making precise Stevens' intuition that we can recover surface shape from the deformed image of a planar curve. This information is robust in that it does not require accurate measurements or the exact details of viewer motion.

These ideas are developed in the Chapter 5 where it is shown that it is possible to recover useful shape and motion information directly from simple properties of the image without going through the computationally difficult and error sensitive process of measuring the exact image velocities or disparities and trying to recover the exact surface shape and 3D viewer motion.

# Chapter 5

# Orientation and Time to Contact from Image Divergence and Deformation

## 5.1 Introduction

Relative motion between an observer and a scene induces deformation in image detail and shape. If these changes are smooth they can be economically described locally by the first order differential invariants of the image velocity field [123] – the curl (vorticity), divergence (dilatation), and shear (deformation) components. The virtue of these invariants is that they have geometrical meaning which does not depend on the particular choice of co-ordinate system. Moreover they are related to the three dimensional structure of the scene and the viewer's motion – in particular the surface orientation and the time to contact [1] – in a simple geometrically intuitive way. Better still, the divergence and deformation components of the image velocity field are unaffected by arbitrary viewer rotations about the viewer centre. They therefore provide an efficient, reliable way of recovering these parameters.

Although the analysis of the differential invariants of the image velocity field has attracted considerable attention [123, 116] their application to real tasks requiring visual inferences has been disappointingly limited [163, 81]. This is because existing methods have failed to deliver reliable estimates of the differential invariants when applied to real images. They have attempted the recovery of dense image velocity fields [47] or the accurate extraction of points or corner features [116]. Both methods have attendant problems concerning accuracy and numerical stability. An additional problem concerns the domain of applications to which estimates of differential invariants can be usefully applied. First order invariants of the image velocity field at a single point in the image cannot be used to provide a *complete* description of shape and motion as attempted in numerous structure from motion algorithms [201]. This in fact requires second order spatial derivatives of the image velocity field [138, 210]. Their power lies in their ability to efficiently recover reliable but incomplete (*partial*) solutions to

---

[1]The time duration before the observer and object collide if they continue with the same relative translational motion [86, 133]

the structure from motion problem. They are especially suited to the domain of active vision, where the viewer makes deliberate (although sometimes imprecise) motions, or in stereo vision, where the relative positions of the two cameras (eyes) are constrained while the cameras (eyes) are free to make arbitrary rotations (eye movements). This study shows that in many cases the extraction of the differential invariants of the image velocity field when augmented with other information or constraints is sufficient to accomplish useful visual tasks.

This chapter begins with a criticism of existing structure from motion algorithms. This motivates the use of partial, incomplete but more reliable solutions to the structure from motion problem. The extraction of the differential invariants of the image velocity field by an active observer is proposed under this framework. Invariants and their relationship to viewer motion and surface shape are then reviewed in detail in sections 5.3.1 and 5.3.2.

The original contribution of this chapter is then introduced in section 5.4 where a novel method to measure the differential invariants of the image velocity field robustly by computing average values from the integral of simple functions of the normal image velocities around image contours is described. This avoids having to recover a dense image velocity field and taking partial derivatives. It also does not require point or line correspondences. Moreover integration provides some immunity to image measurement noise.

In section 5.5 it is shown how an *active* observer making small, deliberate motions can use the estimates of the divergence and deformation of the image velocity field to determine the object surface orientation and time to impact. The results of preliminary real-time experiments in which arbitrary image shapes are tracked using B-spline snakes (introduced in Chapter 3) are presented. The invariants are computed efficiently as closed-form functions of the B-spline snake control points. This information is used to guide a robot manipulator in obstacle collision avoidance, object manipulation and navigation.

## 5.2   Structure from motion

### 5.2.1   Background

The way appearances change in the image due to relative motion between the viewer and the scene is a well known cue for the perception of 3D shape and motion. Psychophysical investigations in the study of the human visual system have shown that visual motion can give vivid 3D impressions. It is called the kinetic depth effect or kineopsis [86, 206].

The computational nature of the problem has attracted considerable attention [201]. Attempts to quantify the perception of 3D shape have determined the number of points and the number of views needed to recover the spatial configuration of the points and the motion compatible with the views. Ullman, in his well-known structure from motion theorem [201], showed that a minimum of three distinct orthographic views of four non-planar points in a *rigid* configuration allow the structure and motion to be completely determined. If perspective projection is assumed two views are, in principle, sufficient. In fact two views of eight points allow the problem to be solved with linear methods [135] while five points from two views give a finite number of solutions [73]. [2]

## 5.2.2   Problems with this approach

The emphasis of these algorithms and the numerous similar approaches that these spawned was to look at point image velocities (or disparities in the discrete motion case) at a number of points in the image, assume *rigidity*, and write out a set of equations relating image velocities to viewer motion. The problem is then mathematically tractable, having been reduced in this way to the solution of a set of equations. Problems of uniqueness and minimum numbers of views and configurations have consequently received a lot of attention in the literature  [136, 73]. This structure from motion approach is however deceivingly simple. Although it has been successfully applied in photogrammetry and some robotics systems [93] when a wide field of view, a large range in depths and a large number of accurately measured image data points are assured, these algorithms have been of little or no practical use in analysing imagery in which the object of interest occupies a small part of the field of view or is distant. This is because the effects due to perspective are often small in practice. As a consequence, the solutions to the perspective structure from motion algorithms are extremely ill-conditioned, often failing in a graceless fashion [197, 214, 60] in the presence of image measurement noise when the conditions listed above are violated. In such cases the effects in the image of viewer translations parallel to the image plane are very difficult to discern from rotations about axes parallel to the image plane.

Another related problem is the *bas–relief* ambiguity [95] in interpreting image velocities when perspective effects are small. In addition to the *speed–scale* ambiguity[3], more subtle effects such as the *bas–relief* problem are not imme-

---

[2] Although these results were publicised in the computer vision literature by Ullman (1979), Longuet-Higgins (1981) and Faugeras and Maybank (1989) they were in fact well known to projective geometers and photogrammetrists in the last century. In particular, solutions were proposed by Chasle (1855); Hesse (1863) (who derived a similar algorithm to Longuet–Higgins's 8-point algorithm); Sturm (1869) (who analysed the case of 5 to 7 points in 2 views); Finsterwalder (1897) and Kruppa (1913) (who applied the techniques to photographs for surveying purposes, showed how to recover the geometry of a scene with 5 points and investigated the finite number of solutions) See [43, 151] for references.

[3] This is obvious from the formulations described above since translational velocities and depths appear together in all terms in the structure from motion equations.

diately evident in these formulations. The bas–relief ambiguity concerns the difficulty of distinguishing between a "shallow" structure close to the viewer and "deep" structures further away. Note that this concerns surface orientation and its effect – unlike the speed–scale ambiguity – is to distort the shape. People experience the same difficulty. We are rather poor at distinguishing a *relief* copy from the same sculpture in the round unless allowed to take a sideways look [121].

Finally these approaches place a lot of emphasis on global rigidity. Despite this it is well known that two (even orthographic) views give vivid 3D impressions even in the presence of a degree of non-rigidity such as the class of smooth transformations e.g. bending transformations which are locally rigid [131].

## 5.2.3   The advantages of partial solutions

The complete solution to the structure from motion problem aims to make explicit quantitative values of the viewer motion (translation and rotation) and then to reconstruct a Euclidean copy of the scene. If these algorithms were made to work successfully, this information could of course be used in a variety of tasks that demand visual information including shape description, obstacle and collision avoidance, object manipulation, navigation and image stabilisation.

Complete solutions to the structure from motion problem are often, in practice, extremely difficult, cumbersome and numerically ill-conditioned. The latter arises because many configurations lead to families of solutions,e.g. the bas–relief problem when perspective effects are small. Also it is not evident that making explicit viewer motion (in particular viewer rotations which give no shape information) and exact quantitative depths leads to useful representations when we consider the purpose of the computation (examples listed above). Not all visual knowledge needs to be of such a precise, quantitative nature. It is possible to accomplish many visual tasks with only *partial* solutions to the structure from motion problem, expressing shape in terms of more qualitative descriptions of shape such as spatial order (relative depths) and *affine structure* (Euclidean shape up to an arbitrary affine transformation or "shear" [130, 131]). The latter are sometimes sufficient, especially if they can be obtained quickly, cheaply and reliably or if they can be augmented with other partial solutions.

In structure from motion two major contributions to this approach have been made in the literature. These include the pioneering work of Koenderink and van Doorn [123, 130], who showed that by looking at the local variation of velocities – rather than point image velocities – useful shape information can be inferred. Although a complete solution can be obtained from second-order derivatives, a more reliable, partial solution can be obtained from certain combinations of first-order derivatives – the divergence and deformation.

More recently, alternative approaches to structure from motion algorithms have been proposed by Koenderink and Van Doorn [131] and Sparr and Nielsen [187]. In the Koenderink and Van Doorn approach, a *weak perspective* projection model and the image motion of three points are used to completely define the affine transformation between the images of the plane defined by the three points. The deviation of a fourth point from this affine transformation specifies shape. Again this is different to the 3D Euclidean shape output by conventional methods. Koenderink shows that it is, however, related to the latter by a *relief* transformation. They show how additional information from extra views can augment this partial solution into a complete solution. This is related to an earlier result by Longuet-Higgins [137], which showed how the velocity of a fourth point relative to the triangle formed by another three provides a useful constraint on translational motion and hence shape. This is also part of a recurrent theme in this thesis that relative local velocity or disparity measurements are reliable geometric cues to shape and motion.

In summary, the emphasis of these methods is to present partial, incomplete but geometrically intuitive solutions to shape recovery from structure from motion.

## 5.3 Differential invariants of the image velocity field

Differential invariants of the image velocity field have been treated by a number of authors. Sections 5.3.1 and 5.3.2 review the main results which were presented originally by Koenderink and Van Doorn [123, 124, 121] in the context of computational vision and the analysis of visual motion. This serves to introduce the notation required for later sections and to clarify some of the ideas presented in the literature.

### 5.3.1 Review

The image velocity of a point in space due to relative motion between the observer and the scene is given by

$$q_t = \frac{(\mathbf{U} \wedge \mathbf{q}) \wedge \mathbf{q}}{\lambda} - \mathbf{\Omega} \wedge \mathbf{q}. \tag{5.1}$$

where $\mathbf{U}$ is the translational velocity, $\mathbf{\Omega}$ is the rotational velocity around the viewer centre and $\lambda$ is the distance to the point. The image velocity consists of two components. The first component is determined by relative translational velocity and encodes the structure of the scene, $\lambda$. The second component depends only on rotational motion about the viewer centre (eye movements). It

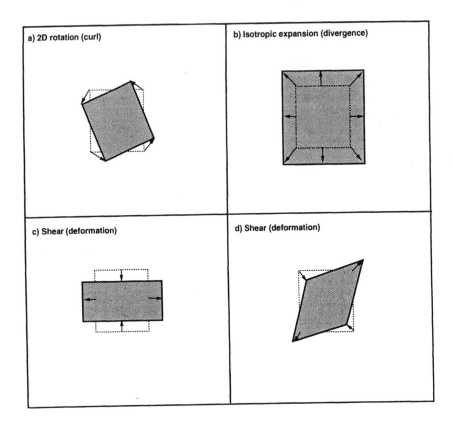

Figure 5.1: Differential invariants of the image velocity field.

*To first order the image velocity field can be decomposed into curl (vorticity), divergence (dilation) and pure shear (deformation) components. The curl, divergence and the magnitude of the deformation are differential invariants and do not depend on the choice of image co-ordinate system. Their effect on apparent image shape can be described by four independent components of an affine transformation. These are: (a) a 2D rotation; (b) an isotropic expansion (scaling); (c) and (d) two deformation components. The latter two are both pure shears about different axes. Any deformation can be conveniently decomposed into these two components. Each component is dependent on an arbitrary choice of co-ordinate system and is not a differential invariants.*

gives no useful information about the depth of the point or the shape of the visible surface. It is this rotational component which complicates the interpretation of visual motion. The effects of rotation are hard to extricate however, although numerous solutions have been proposed [150]. As a consequence, point image velocities and disparities do not encode shape in a simple efficient way since the rotational component is often arbitrarily chosen to shift attention and gaze by camera rotations or eye movements.

We now look at the local variation of image velocities in the vicinity of the ray $\mathbf{q}$. Consider an arbitrary co-ordinate system (the final results will be invariant to this choice) with the $x - y$ plane spanning the image plane (tangent plane of projection sphere at $\mathbf{q}$) and the $z$-axis aligned with the ray. In this co-ordinate system the translational velocity has components $\{U_1, U_2, U_3\}$ and the angular velocity has components $\{\Omega_1, \Omega_2, \Omega_3\}$. Let the image velocity field at a point $(x, y)$ in the vicinity of $\mathbf{q}$ be represented as a 2D vector field, $\vec{v}(x, y)$ with $x$ and $y$ components $(u, v)$.

For a sufficiently small field of view or for a small neighbourhood (defined more precisely below), the image velocity field can be described by a translation in the image $(u_0, v_0)$ and by the first-order partial derivatives of the image velocity $(u_x, u_y, v_x, v_y)$, where [210, 150]:

$$u_0 = -\frac{U_1}{\lambda} - \Omega_2 \tag{5.2}$$

$$v_0 = -\frac{U_2}{\lambda} + \Omega_1 \tag{5.3}$$

$$u_x = \frac{U_3}{\lambda} + \frac{U_1 \lambda_x}{\lambda^2} \tag{5.4}$$

$$u_y = +\Omega_3 + \frac{U_1 \lambda_y}{\lambda^2} \tag{5.5}$$

$$v_x = -\Omega_3 + \frac{U_2 \lambda_x}{\lambda^2} \tag{5.6}$$

$$v_y = \frac{U_3}{\lambda} + \frac{U_2 \lambda_y}{\lambda^2} \tag{5.7}$$

and where the $x$ and $y$ subscripts represent differentiation with respect to these spatial parameters. Note that there are six equations in terms of the eight unknowns of viewer motion and surface orientation. The system of equations is thus under-constrained.

An image feature or shape will experience a transformation as a result of the image velocity field. The transformation from a shape at time $t$ to the deformed shape at a small instant of time later, at $t + \delta t$, can also be approximated by a linear transformation – an *affine* transformation. In fact, any arbitrary small smooth transformation is linear in the limit and well approximated by the first derivative in a sufficiently small region.

To first order the image velocity field at a point $(x, y)$ in the neighbourhood of **q** can be approximated by:

$$\begin{bmatrix} u \\ v \end{bmatrix} = \begin{bmatrix} u_0 \\ v_0 \end{bmatrix} + \begin{bmatrix} u_x & u_y \\ v_x & v_y \end{bmatrix} \begin{bmatrix} x \\ y \end{bmatrix} + O(x^2, xy, y^2) \qquad (5.8)$$

where $O(x^2, xy, y^2)$ represents non-linear terms which are neglected in this analysis. The first term is a vector $[u_0, v_0]$ representing a pure translation while the second term is a $2 \times 2$ tensor – the velocity gradient tensor – and represents the distortion of the image shape.

We can decompose the velocity gradient tensor into three components, where each term has a simple geometric significance invariant under the transformation of the image co-ordinate system. [4] These components are the first-order differential invariants of the image velocity field – the vorticity (curl) , dilatation (divergence) and pure shear (deformation) components.

$$
\begin{aligned}
\begin{bmatrix} u_x & u_y \\ v_x & v_y \end{bmatrix} &= \frac{\text{curl}\vec{v}}{2} \begin{bmatrix} 0 & -1 \\ 1 & 0 \end{bmatrix} + \frac{\text{div}\vec{v}}{2} \begin{bmatrix} 1 & 0 \\ 0 & 1 \end{bmatrix} + \\
&\quad \frac{\text{def}\vec{v}}{2} \begin{bmatrix} \cos\mu & -\sin\mu \\ \sin\mu & \cos\mu \end{bmatrix} \begin{bmatrix} 1 & 0 \\ 0 & -1 \end{bmatrix} \begin{bmatrix} \cos\mu & \sin\mu \\ -\sin\mu & \cos\mu \end{bmatrix} \quad (5.9) \\
&= \frac{\text{curl}\vec{v}}{2} \begin{bmatrix} 0 & -1 \\ 1 & 0 \end{bmatrix} + \frac{\text{div}\vec{v}}{2} \begin{bmatrix} 1 & 0 \\ 0 & 1 \end{bmatrix} + \frac{\text{def}\vec{v}}{2} \begin{bmatrix} \cos 2\mu & \sin 2\mu \\ \sin 2\mu & -\cos 2\mu \end{bmatrix}
\end{aligned}
$$

where $\text{curl}\vec{v}$, $\text{div}\vec{v}$ and $\text{def}\vec{v}$ represent the curl, divergence and deformation components and where $\mu$ specifies the orientation of the axis of expansion (maximum extension). [5] These quantities are defined by:

$$\text{div}\vec{v} = (u_x + v_y) \qquad (5.10)$$
$$\text{curl}\vec{v} = -(u_y - v_x) \qquad (5.11)$$
$$(\text{def}\vec{v})\cos 2\mu = (u_x - v_y) \qquad (5.12)$$
$$(\text{def}\vec{v})\sin 2\mu = (u_y + v_x). \qquad (5.13)$$

These can be derived in terms of differential invariants [116] or can be simply considered as combinations of the partial derivatives of the image velocity field with simple geometric meanings. The curl, divergence and the magnitude of the deformation are scalar invariants and do not depend on the particular choice of co-ordinate system. The axes of maximum extension and contraction rotate with rotations of the image plane axes.

---

[4] The decomposition is known in applied mechanics as the Cauchy–Stokes decomposition theorem [5].

[5] $(\cos\mu, \sin\mu)$ is the eigenvector of the traceless and symmetric component of the velocity tensor. It corresponds the positive eigenvalue with magnitude $\text{def}\vec{v}$. The other eigenvector specifies the axis of contraction and is orthogonal. It corresponds to the negative eigenvalue with magnitude $-\text{def}\vec{v}$.

Consider the effect of these components on the transformation of apparent image shapes (figure 5.1). The curl component that measures the 2D rigid rotation or change in orientation of patches in the image. The divergence term specifies scale or size changes. The deformation term specifies the distortion of the image shape as a shear (expansion in a specified direction with contraction in a perpendicular direction in such a way that area is unchanged). It is specified by an axis of expansion and a magnitude (the size of the change in this direction).

It will be seen below that the main advantage of considering the differential invariants of the image velocity field is that the deformation component efficiently encodes the orientation of the surface while the divergence component can be used to provide an estimate of the time to contact or collision.

Before looking at the 3D interpretation of these invariants, it is important to make explicit under which conditions it is reasonable to consider the image velocity field to be well approximated by its first order terms. This requires that the transformation is locally equivalent to an affine transformation. For example, parallel lines must remain parallel or equivalently the transformation from a plane in the world to the image plane must also be described by an affine mapping. This is known as *weak* perspective. By inspecting the quadratic terms in the equation of the image velocity in the vicinity of a point in the image (5.1) it is easy to show that we require in the field of interest:

$$\frac{\Delta\lambda}{\lambda} \ll 1 \tag{5.14}$$

$$\frac{\Omega.\delta}{\Omega.q} \ll 1 \tag{5.15}$$

where $\delta$ is a difference between two ray directions and defines the field of view in radians and $\Delta\lambda$ is the depth of the relief in the field of view. A useful empirical result is that if the distance to the object is greater than the depth of the relief by an order of magnitude [193] then the assumption of weak perspective is a good approximation to perspective projection.

At close distances "looming" or "fanning" effects will become noticeable and the affine transformation is insufficient to describe the changes in the image. In many practical cases, however, it is possible to restrict attention to small fields of view in which the weak perspective model is valid.

## 5.3.2 Relation to 3D shape and viewer ego-motion

The relationships between the observed differential invariants and the three-dimensional configuration and the viewer motion are given. In particular the differential invariants are expressed in terms of the viewer translation $(U_1/\lambda, U_2/\lambda, U_3/\lambda)$

and the surface orientation $(\lambda_x/\lambda, \lambda_y/\lambda)$. From (5.2) to (5.13) we have:

$$u_0 = -\frac{U_1}{\lambda} - \Omega_2$$

$$v_0 = -\frac{U_2}{\lambda} + \Omega_1$$

$$\text{curl}\,\vec{v} = -2\Omega_3 + \frac{(-U_1\lambda_y + U_2\lambda_x)}{\lambda^2} \qquad (5.16)$$

$$\text{div}\,\vec{v} = 2\frac{U_3}{\lambda} + \frac{U_1\lambda_x + U_2\lambda_y}{\lambda^2} \qquad (5.17)$$

$$(\text{def}\,\vec{v})\cos 2\mu = \frac{(U_1\lambda_x - U_2\lambda_y)}{\lambda^2} \qquad (5.18)$$

$$(\text{def}\,\vec{v})\sin 2\mu = \frac{(U_1\lambda_y + U_2\lambda_x)}{\lambda^2}. \qquad (5.19)$$

Note that the average image translation $(u_0, v_0)$ can always be cancelled out by appropriate camera rotations (eye movements) $(\Omega_1, \Omega_2)$. Also note that divergence and deformation are unaffected by viewer rotations such as panning or tilting of the camera or eye movements whereas these could lead to considerable changes in point image velocities or disparities.

The differential invariants depend on the viewer motion, depth and surface orientation. We can express them in a co-ordinate free manner by introducing two 2D vector quantities: the component of translational velocity parallel to the image plane scaled by depth, $\lambda$, $\mathbf{A}$ where:

$$\mathbf{A} = \left(\frac{U_1}{\lambda}, \frac{U_2}{\lambda}\right)$$

$$= \frac{\mathbf{U} - (\mathbf{U}.\mathbf{q})\mathbf{q}}{\lambda} \qquad (5.20)$$

and the *depth gradient* scaled by depth [6], $\mathbf{F}$, to represent the surface orientation and which we define in terms of the 2D vector gradient: [7]

$$\mathbf{F} = \left(\frac{\lambda_x}{\lambda}, \frac{\lambda_y}{\lambda}\right) \qquad (5.21)$$

$$= \frac{\text{grad}\lambda}{\lambda}. \qquad (5.22)$$

---

[6] Koenderink [121] defines $\mathbf{F}$ as a "nearness gradient" – $\text{grad}(log(1/\lambda))$. In this section $\mathbf{F}$ is defined as a scaled depth gradient. These two quantities differ by a sign.

[7] There are three simple ways to represent surface orientation: components of a unit vector, $\mathbf{n}$; gradient space representation $(p, q)$ and the spherical co-ordinates $(\sigma, \tau)$. Changing from one representation to another is trivial and is listed here for completeness.

$$\tan\sigma = \sqrt{p^2 + q^2}$$

$$\tan\tau = \frac{q}{p}$$

$$\mathbf{n} = (\sin\sigma\cos\tau, \sin\sigma\sin\tau, \cos\sigma).$$

The magnitude of the depth gradient determines the tangent of the *slant* of the surface (angle between the surface normal and the visual direction). It vanishes for a frontal view and is infinite when the viewer is in the tangent plane of the surface. Its direction specifies the direction in the image of increasing distance. This is equal to the *tilt* of the surface tangent plane, $\tau$. The exact relationship between the magnitude and direction of $\mathbf{F}$ and the slant and tilt of the surface $(\sigma, \tau)$ is given by:

$$|\mathbf{F}| = \tan \sigma \tag{5.23}$$

$$\angle \mathbf{F} = \tau \tag{5.24}$$

With this new notation equations (5.16, 5.17, 5.18 and 5.19) can be re-written to show the relation between the differential invariants, the motion parameters and the surface position and orientation:

$$\text{curl}\vec{v} = -2\mathbf{\Omega}.\mathbf{q} + \mathbf{F} \wedge \mathbf{A} \tag{5.25}$$

$$\text{div}\vec{v} = \frac{2\mathbf{U}.\mathbf{q}}{\lambda} + \mathbf{F}.\mathbf{A} \tag{5.26}$$

$$\text{def}\vec{v} = |\mathbf{F}||\mathbf{A}| \tag{5.27}$$

where $\mu$ (which specifies the axis of maximum extension) bisects $\mathbf{A}$ and $\mathbf{F}$:

$$\mu = \frac{\angle \mathbf{A} + \angle \mathbf{F}}{2}. \tag{5.28}$$

The geometric significance of these equations is easily seen with a few examples (see below). Note that this formulation clearly exposes both the speed–scale ambiguity – translational velocities appear scaled by depth making it impossible to determine whether the effects are due to a nearby object moving slowly or a far-away object moving quickly – and the bas–relief ambiguity. The latter manifests itself in the appearance of surface orientation, $\mathbf{F}$, with $\mathbf{A}$. Increasing the slant of the surface $\mathbf{F}$ while scaling the movement by the same amount will leave the local image velocity field unchanged. Thus, from two weak perspective views and with no knowledge of the viewer translation, it is impossible to determine whether the deformation in the image is due to a large $|\mathbf{A}|$ (large "turn" of the object or "vergence angle") and a small slant or a large slant and a small rotation around the object. Equivalently a nearby "shallow" object will produce the same effect as a far away "deep" structure. We can only recover the depth gradient $\mathbf{F}$ up to an unknown scale. These ambiguities are clearly exposed with this analysis whereas this insight is sometimes lost in the purely algorithmic approaches to solving the equations of motion from the observed point image velocities.

It is interesting to note the similarity between the equations of motion parallax (introduced in Chapter 2 and listed below for the convenience of comparison) which relate the relative image velocity between two nearby points, $\mathbf{q}^{(2)} - \mathbf{q}^{(1)}$, to their relative inverse depths:

$$\mathbf{q}_t^{(2)} - \mathbf{q}_t^{(1)} = [(\mathbf{U} \wedge \mathbf{q}) \wedge \mathbf{q}] \left[ \frac{1}{\lambda^{(2)}} - \frac{1}{\lambda^{(1)}} \right] \tag{5.29}$$

and the equation relating image deformation to surface orientation:

$$\mathrm{def}\,\vec{\mathbf{v}} = |(\mathbf{U} \wedge \mathbf{q}) \wedge \mathbf{q}| \left| \left[ \mathbf{grad}(\frac{1}{\lambda}) \right] \right| . \tag{5.30}$$

The results are essentially the same, relating local measurements of relative image velocities to scene structure in a simple way which is uncorrupted by the rotational image velocity component. In the first case (5.29), the depths are discontinuous and differences of discrete velocities are related to the difference of inverse depths. In the latter case, (5.30), the surface is assumed smooth and continuous and derivatives of image velocities are related to derivatives of inverse depth.

Some examples on real image sequences are considered. These highlight the effect of viewer motion and surface orientation on the observed image deformations.

1. Panning and tilting $(\Omega_1, \Omega_2)$ of the camera has no effect locally on the differential invariants (5.2). They just shift the image. At any moment eye movements can locally cancel the effect of the mean translation. This is the purpose of fixation.

2. A rotation about the line of sight leads to an opposite rotation in the image (curl, (5.25)). This is simply a 2D rigid rotation.

3. A translation towards the surface patch (figure 5.2a and b) leads to a uniform expansion in the image, i.e. a positive divergence. This encodes distance in temporal units, i.e. as a time to contact or collision. Both rotations about the ray and translations along the ray produce no deformation in image detail and hence contain no information about the surface orientation.

4. Deformation arises for translational motion perpendicular to the visual direction. The magnitude and axes of the deformation depend on the orientation of the surface and the direction of translation. Figure 5.2 shows a surface slanted away from the viewer but with zero tilt, i.e. the depth increases as we move horizontally from left to right. Figure 5.2c shows the image after a sideways movement to the left with a camera rotation to keep the target in the centre of the field of view. The divergence and deformation components are immediately evident. The contour shape extends

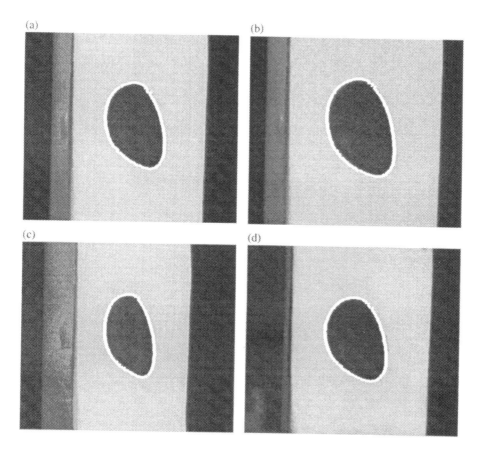

Figure 5.2: Distortions in apparent shape due to viewer motion.

*(a) The image of a planar contour (zero tilt and positive slant, i.e. the direction of increasing depth, **F**, is horizontal and from left to right). The image contour is localised automatically by a B-spline snake initialised in the centre of the field of view. (b) The effect on apparent shape of a viewer translation towards the target. The shape undergoes an isotropic expansion (positive divergence). (c) The effect on apparent shape when the viewer translates to the left while fixating on the target (i.e. **A** is horizontal, right to left). The apparent shape undergoes an isotropic contraction (negative divergence which reduces the area) and a deformation in which the axis of expansion is vertical. These effects are predicted by equations (5.25, 5.26, 5.27 and 5.28) since the bisector of the direction of translation and the depth gradient is the vertical. (d) The opposite effect when the viewer translates to the right. The axes of contraction and expansion are reversed. The divergence is positive. Again the curl component vanishes.*

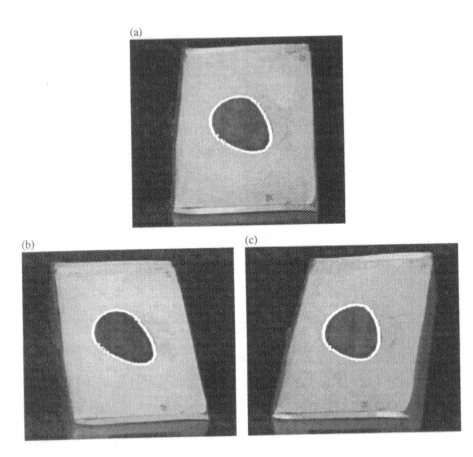

Figure 5.3: Image deformations and rotations due to viewer motion.

*(a) The image of a planar contour (90° tilt and positive slant – i.e. the direction of increasing depth, **F**, is vertical, bottom to top). (b) The effect on apparent shape of a viewer translation to the left. The contour undergoes a deformation with the axis of expansion at 135° to the horizontal. The area of the contour is conserved (vanishing divergence). The net rotation is however non-zero. This is difficult to see from the contour alone. It is obvious, however, by inspection of the sides of the box, that there has been a net clockwise rotation. (c) These effects are reversed when the viewer translates to the right.*

along the vertical axis and contracts along the horizontal as predicted by equations (5.28). This is followed by a reduction in apparent size due to the foreshortening effect as predicted by (5.26). This result is intuitively obvious since a movement to the left makes the object appear in a less frontal view. From (5.25) we see that the curl component vanishes. There is no rotation of the image shape. Movement to the right (figure 5.2d) reverses these effects.

5. For sideways motion with a surface with non-zero tilt relative to direction of translation, the axis of contraction and expansion are no longer aligned with the image axes. Figure 5.3 shows a surface whose tilt is $90^o$ (depth increases as we move vertically in the image). A movement to the left with fixation causes a deformation. The vertical velocity gradient is immediately apparent. The axis of expansion of the deformation is at $135^o$ to the left–right horizontal axis, again bisecting **F** and **A**. There is no change in the area of the shape (zero divergence) but a clockwise rotation. The evidence for the latter is that the horizontal edges have remained horizontal. A pure deformation alone would have changed these orientations. The curl component has the effect of nulling the net rotation. If the direction of motion is reversed the axis of expansion moves to $45^o$ as predicted. Again the basic equations of (5.25, 5.26, 5.27 and 5.28) adequately describe these effects.

## 5.3.3   Applications

Applications of estimates of the differential invariants of the image velocity field are summarised below. It has already been noted that measurement of the differential invariants in a single neighbourhood is insufficient to completely solve for the structure and motion since we have six equations in the eight unknowns of scene structure and motion. In a single neighbourhood a complete solution would require the computation of second order derivatives [138, 210] to generate sufficient equations to solve for the unknowns. Even then solution of the resulting set of non-linear equations is non-trivial.

In the following, the information available from the first-order differential invariants alone is investigated. It will be seen that the differential invariants are usually sufficient to perform useful visual tasks with the added benefit of being geometrically intuitive. Useful applications include providing information which is used by pilots when landing aircraft [86], estimating time to contact in braking reactions [133] and in the recovery of 3D shape up to a relief transformation [130, 131].

1. **With knowledge of translation but arbitrary rotation**

   An estimate of the direction of translation is usually available when the viewer is making deliberate movements (in the case of active vision) or in the case of binocular vision (where the camera or eye positions are constrained). It can also be estimated from image measurements by motion parallax [138, 182].

   If the viewer translation is known, equations (5.27), (5.28) and (5.26) are sufficient to unambiguously recover the surface orientation and the distance to the object in temporal units. Due to the speed–scale ambiguity the latter is expressed as a time to contact. A solution can be obtained in the following way.

   - The axis of expansion ($\mu$) of the deformation component and the projection in the image of the direction of translation ($\angle\mathbf{A}$) allow the recovery of the tilt of the surface (5.28).
   - We can then subtract the contribution due to the surface orientation and viewer translation parallel to the image axis from the image divergence (5.26). This is equal to $|\mathrm{def}\,\bar{\mathbf{v}}|\cos(\tau - \angle\mathbf{A})$. The remaining component of divergence is due to movement towards or away from the object. This can be used to recover the time to contact, $t^c$:

$$t^c = \frac{\lambda}{\mathbf{U}.\mathbf{q}}. \tag{5.31}$$

   This has been recovered despite the fact that the viewer translation may not be parallel to the visual direction.

   - The time to contact fixes the viewer translation in temporal units. It allows the specification of the magnitude of the translation parallel to the image plane (up to the same speed–scale ambiguity), $\mathbf{A}$. The magnitude of the deformation can then be used to recover the slant, $\sigma$, of the surface from (5.27).

   The advantage of this formulation is that camera rotations do not affect the estimation of shape and distance. The effects of errors in the direction of translation are clearly evident as scalings in depth or by a relief transformation [121].

2. **With fixation**

   If the cameras or eyes rotate to keep the object of interest in the middle of the image (null the effect of image translation) the eight unknowns are reduced to six. The magnitude of the rotations needed to bring the object back to the centre of the image determines $\mathbf{A}$ and hence allows us to solve for these unknowns, as above. Again the major effect of any error in the estimate of rotation is to scale depth and orientations.

3. **With no additional information – constraints on motion**

Even without any additional assumptions it is still possible to obtain useful information from the first-order differential invariants. The information obtained is best expressed as bounds. For example inspection of equation (5.26) and (5.27) shows that the time to contact must lie in an interval given by:

$$\frac{1}{t^c} = \frac{\mathrm{div}\,\vec{v}}{2} \pm \frac{\mathrm{def}\,\vec{v}}{2}. \tag{5.32}$$

The upper bound on time to contact occurs when the component of viewer translation parallel to the image plane is in the opposite direction to the depth gradient. The lower bound occurs when the translation is parallel to the depth gradient. The upper and lower estimates of time to contact are equal when there is no deformation component. This is the case in which the viewer translation is along the ray or when viewing a fronto-parallel surface (zero depth gradient locally). The estimate of time to contact is then exact. A similar equation was recently described by Subbarao [189]. He describes the other obvious result that knowledge of the curl and deformation components can be used to estimate bounds on the rotational component about the ray,

$$\Omega.\mathbf{q} = -\frac{\mathrm{curl}\,\vec{v}}{2} \pm \frac{\mathrm{def}\,\vec{v}}{2}. \tag{5.33}$$

4. **With no additional information – the constraints on 3D shape**

Koenderink and Van Doorn [130] showed that surface shape information can be obtained by considering the variation of the deformation component alone in small field of view when weak perspective is a valid approximation. This allows the recovery of 3D shape up to a scale and relief transformation. That is they effectively recover the axis of rotation of the object but not the magnitude of the turn. This yields a family of solution depending on the magnitude of the turn. Fixing the latter determines the slants and tilts of the surface. This has recently been extended in the affine structure from motion theorem [131, 187].

The invariants of the image velocity field encode the relations between shape and motion in a concise, geometrically appealing way. Their measurement and application to real examples requiring action on visual inferences will now be discussed.

## 5.3.4 Extraction of differential invariants

The analysis above treated the differential invariants as observables of the image. There are a number of ways of extracting the differential invariants from the

image. These are summarised below and a novel method based on the moments
of areas enclosed by closed curves is presented.

1. **Partial derivative of image velocity field**

   This is the most commonly stressed approach. It is based on recovering a
   dense field of image velocities and computing the partial derivatives using
   discrete approximation to derivatives [126] or a least squares estimation of
   the affine transformation parameters from the image velocities estimated
   by spatio-temporal methods [163, 47]. The recovery of the image velocity
   field is usually computationally expensive and ill-conditioned.

2. **Point velocities in a small neighbourhood**

   The image velocities of a minimum of three points in a small neighbourhood are sufficient, in principle, to estimate the components of the affine
   transformation and hence the differential invariants [116, 130]. In fact it
   is only necessary to measure the change in area of the triangle formed
   by the three points and the orientations of its sides. However this is the
   minimum information. There is no redundancy in the data and hence
   this requires very accurate image positions and velocities. In [53] this is
   attempted by tracking large numbers of "corner" features [97, 208] and using Delaunay triangulation [33] in the image to approximate the physical
   world by planar facets. Preliminary results showed that the localisation of
   "corner" features was insufficient for reliable estimation of the differential
   invariants.

3. **Relative orientation of line segments**

   Koenderink [121] showed how temporal texture density changes can yield
   estimates of the divergence. He also presented a method for recovering
   the curl and shear components that employs the orientations of texture
   elements.

   From (5.10) it is easy to show that the change in orientation (clockwise),
   $\Delta\phi$ of an element with orientation $\phi$ is given to first order by [124]

   $$\Delta\phi = -\frac{\text{curl}\vec{v}}{2} + \frac{1}{2}\text{def}\vec{v}\sin 2(\phi - \mu). \tag{5.34}$$

   Orientations are not affected by the divergence term. They are only affected by the curl and deformation components. In particular the curl
   component changes all the orientations by the same amount. It does not
   affect the angles between the image edges. These are only affected by the
   deformation component. The relative changes in orientation can be used to
   recover deformation in a simple way since the effects of the curl component

are cancelled out. By taking the difference of (5.34) for two orientations, $\phi_1$ and $\phi_2$, it is easy to show (using simple trigonometric relations) that the relative change in orientation specifies both the magnitude, $\mathrm{def}\vec{v}$, and axis of expansion of the shear, $\mu$, as shown below.

$$\Delta\phi_2 - \Delta\phi_1 = \mathrm{def}\vec{v}\left[\sin(\phi_2 - \phi_1)\cos 2\left(\frac{\phi_2 + \phi_1}{2} - \mu\right)\right].\qquad(5.35)$$

Measurement at three oriented line segments is sufficient to completely specify the deformation components. Note that the recovery of deformation can be done without any explicit co-ordinate system and even without a reference orientation. The main advantage is that point velocities or partial derivatives are not required. Koenderink proposes this method as being well suited for implementation in a physiological setting [121].

## 4. Curves and closed contours

We have seen how to estimate the differential invariants from point and line correspondences. Sometimes these are not available or are poorly localised. Often we can only reliably extract portions of curves (although we can not always rely on the end points) or closed contours.

Image shapes or contours only "sample" the image velocity field. At contour edges it is only possible to measure the normal component of image velocity. This information can in certain cases be used to recover the image velocity field. Waxman and Wohn [211] showed how to recover the full velocity field from the normal components of image contours. In principle, measurement of eight normal velocities around a contour allow the characterisation of the full velocity field for a planar surface. Kanatani [115] also relates line integrals of image velocities around closed contours to the motion and orientation parameters of a planar contour. We will not attempt to solve for these parameters directly but only to recover the divergence and deformation.

In the next section, we analyse the changing shape of a closed contour (not just samples of normal velocities) to recover the differential invariants. Integral theorems exist which express the average value of the differential invariants in terms of integrals of velocity around boundaries of regions. They deal with averages and not point properties and will potentially have better immunity to noise. Another advantage of closed curves is that point or line correspondences are not required. Only the correspondence of image shapes.

## 5.4    Recovery of differential invariants from closed contours

It has been shown that the differential invariants of the image velocity field conveniently characterise the changes in apparent shape due to relative motion between the viewer and scene. Contours in the image sample this image velocity field. It is usually only possible, however, to recover the normal image velocity component from local measurements at a curve [202, 100]. It is now shown that this information is often sufficient to estimate the differential invariants within closed curves. Moreover, since we are using the integration of normal image velocities around closed contours to compute average values of the differential invariants, this method has a noise-defeating effect leading to reliable estimates.

The approach is based on relating the temporal derivative of the area of a closed contour and its moments to the invariants of the image velocity field. This is a generalisation of the result derived by Maybank [148], in which the rate of change of area scaled by area is used to estimate the divergence of the image velocity field.

The advantage is that it is not necessary to track point features in the image. Only the correspondence between shapes is required. The computationally difficult, ill-conditioned and poorly defined process of making explicit the full image velocity field [100] is avoided. Moreover, areas can be estimated accurately, even when the full set of first order derivatives can not be obtained.

The moments of area of a contour are defined in terms of an area integral with boundaries defined by the contour in the image plane (figure 5.4);

$$I_f = \int_{a(t)} f dx dy \qquad (5.36)$$

where $a(t)$ is the area of a contour of interest at time $t$ and $f$ is a scalar function of image position $(x, y)$ that defines the moment of interest. For instance setting $f = 1$ gives the zero order moment of area (which we label $I_0$). This is simply the area of the contour. Setting $f = x$ or $f = y$ gives the first-order moments about the image $x$ and $y$ axes respectively.

The moments of area can be measured directly from the image (see below for a novel method involving the control points of the B-spline snake). Better still, their temporal derivatives can also be measured. Differentiating (5.36) with

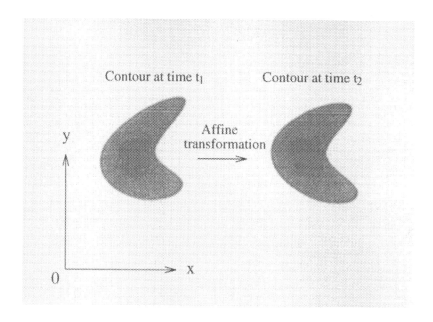

Figure 5.4: The temporal evolution of image contours.
*For small fields of view the distortion in image shape can be described locally by an affine transformation. The components of the affine transformation can be expressed in terms of contour integrals of normal image velocities. More conveniently the temporal derivatives of the area and its moments can be used to characterise the distortion in apparent shape and the affine transformation.*

respect to time and using a result from calculus [61] it can be shown that [8]

$$\frac{d}{dt}(I_f) = \frac{d}{dt}\left[\int_{a(t)} f\, dx\, dy\right] \tag{5.37}$$

$$= \oint_{c(t)} [f\vec{\mathbf{v}}.\mathbf{n}^p]\, ds \tag{5.38}$$

where $\vec{\mathbf{v}}.\mathbf{n}^p$ is the normal component of the image velocity $\vec{\mathbf{v}}$ at a point on the contour. Note that the temporal derivatives of moments of area are simply equivalent to integrating the normal image velocities at the contour weighted by a scalar $f(x,y)$.

By Green's theorem, an integral over the contour $c(t)$, can be re-expressed as an integral over the area enclosed by the contour, $a(t)$. The right-hand side of (5.38) can be re-expressed as:

$$\frac{d}{dt}(I_f) = \int_{a(t)} [\mathrm{div}(f\vec{\mathbf{v}})]\, dxdy \tag{5.39}$$

$$= \int_{a(t)} [f\,\mathrm{div}\vec{\mathbf{v}} + (\vec{\mathbf{v}}.\mathbf{grad}f)]\, dxdy \tag{5.40}$$

$$= \int_{a(t)} [f\,\mathrm{div}\vec{\mathbf{v}} + f_x u + f_y v]\, dxdy \tag{5.41}$$

Assuming that the image velocities can be represented by (5.8) in the area of interest, i.e. by constant partial derivatives:

$$\frac{d}{dt}(I_f) = u_0 \int_{a(t)} [f_x]\, dxdy + u_x \int_{a(t)} [xf_x + f]\, dxdy + u_y \int_{a(t)} [yf_x]\, dxdy \tag{5.42}$$

$$+ v_0 \int_{a(t)} [f_y]\, dxdy + v_x \int_{a(t)} [xf_y + f]\, dxdy + v_y \int_{a(t)} [yf_y + f]\, dxdy.$$

The left hand side is the temporal derivative of the moment of area described by $f$. The integrals on the right-hand side are simply moments of area (which are directly measurable). The coefficients of each term are the required parameters of the affine transformation. The equations are geometrically intuitive. The image velocity field deforms the shape of contours in the image. Shape can be described by moments of area. Hence measuring the change in the moments of area is an alternative to describing the transformation. In this way the change in the moments of area have been expressed in terms of the parameters of the affine transformation.

---

[8] This equation can be derived by considering the *flux* linking the area of the contour. This changes with time since the contour is carried by the velocity field. The *flux* field in our example does not change with time. Similar integrals appear in fluid mechanics, e.g. the *flux transport theorem* [61].

If we initially set up the $x - y$ co-ordinate system at the centroid of the image contour of interest so that the first moments are zero, equation (5.42) with $f = x$ and $f = y$ shows that the centroid of the deformed shape specifies the mean translation $[u_0, v_0]$. Setting $f = 1$ leads to the extremely simple and powerful result that the divergence of the image velocity field can be estimated as the derivative of area scaled by area.

$$\frac{dI_0}{dt} = I_0(u_x + v_y) \tag{5.43}$$

$$\frac{da(t)}{dt} = a(t)\mathrm{div}\,\vec{v}. \tag{5.44}$$

Increasing the order of the moments, i.e. different values of $f(x, y)$, generates new equations and additional constraints. In principle, if it is possible to find six linearly independent equations, we can solve for the affine transformation parameters and combine the co-efficients to recover the differential invariants. The validity of the affine approximation can be checked by looking at the error between the transformed and observed image contours. The choice of which moments to use is a subject for further work. Listed below are some of the simplest equations which have been useful in the experiments presented here.

$$\frac{d}{dt}\begin{bmatrix} I_0 \\ I_x \\ I_y \\ I_{x^2} \\ I_{y^2} \\ I_{x^3y} \end{bmatrix} = \begin{bmatrix} 0 & 0 & I_0 & 0 & 0 & I_0 \\ I_0 & 0 & 2I_x & I_y & 0 & I_x \\ 0 & I_0 & I_y & 0 & I_x & 2I_y \\ 2I_x & 0 & 3I_{x^2} & 2I_{xy} & 0 & I_{x^2} \\ 0 & 2I_y & I_{y^2} & 0 & 2I_{xy} & 3I_{y^2} \\ 3I_{x^2y} & I_{x^3} & 4I_{x^3y} & 3I_{x^2y^2} & I_{x^4} & 2I_{x^3y} \end{bmatrix}\begin{bmatrix} u_0 \\ v_0 \\ u_x \\ u_y \\ v_x \\ v_y \end{bmatrix}. \tag{5.45}$$

(Note that in this equation subscripts are used to label the moments of area. The left-hand side represents the temporal derivative of the moments in the column vector.) In practice certain contours may lead to equations which are not independent or ill-conditioned. The interpretation of this is that the normal components of image velocity are insufficient to recover the true image velocity field globally, e.g. a fronto-parallel circle rotating about the optical axis. This was termed the "aperture problem in the large" by Waxman and Wohn [211] and investigated by Berghom and Carlsson [19]. Note however, that it is always possible to recover the divergence from a closed contour.

## 5.5 Implementation and experimental results

### 5.5.1 Tracking closed loop contours

The implementation and results follow. Multi-span closed loop B-spline snakes (introduced in Chapter 3) are used to localise and track closed image contours.

The B-spline is a curve in the image plane

$$\mathbf{x}(s) = \sum_i f_i(s)\mathbf{Q}_i \qquad (5.46)$$

where $f_i$ are the spline basis functions with coefficients $\mathbf{Q}_i$ (control points of the curve) and $s$ is a curve parameter (not necessarily arc length). The snakes are initialised as points in the centre of the image and are forced to expand radially outwards until they were in the vicinity of an edge where image "forces" make the snake stabilise close to a high contrast closed contour. Subsequent image motion is automatically tracked by the snake.

B-spline snakes have useful properties such as local control and continuity. They also compactly represent image curves. In our applications they have the additional advantage that the area enclosed is a simple function of the control points. This also applies to the other area moments.

From Green's theorem in the plane it is easy to show that the area enclosed by a curve with parameterisation $x(s)$ and $y(s)$ is given by:

$$a = \int_{s_0}^{s_N} x(s)y'(s)ds \qquad (5.47)$$

where $x(s)$ and $y(s)$ are the image plane components of the B-spline and $y'(s)$ is the derivative with respect to the curve parameter $s$.

For a B-spline, substituting (5.46) and its derivative:

$$a(t) = \int_{s_0}^{s_N} \sum_i \sum_j (Q_{x_i}Q_{y_j})f_i f_j' ds \qquad (5.48)$$

$$= \sum_i \sum_j (Q_{x_i}Q_{y_j}) \int_{s_0}^{s_N} f_i f_j' ds. \qquad (5.49)$$

Note that for each span of the B-spline and at each time instant the basis functions remain unchanged. The integrals can thus be computed off-line in closed form. (At most 16 coefficients need be stored. In fact due to symmetry there are only 10 possible values for a cubic B-spline). At each time instant multiplication with the control point positions gives the area enclosed by the contour. This is extremely efficient, giving the exact area enclosed by the contour. The same method can be used for higher moments of area as well.

## 5.5.2   Recovery of time to contact and surface orientation

Here we present the results of a préliminary implementation of the theory. The examples are based on a camera mounted on a robot arm whose translations are deliberate while the rotations around the camera centre are performed to keep the target of interest in the centre of its field of view. The camera intrinsic parameters (image centre, scaling factors and focal length) and orientation are unknown. The direction of translation is assumed known and expressed with bounds due to uncertainty.

Figures 5.5 to 5.10 show the results of these techniques applied to real image sequences from the Adept robot workspace, as well as other laboratory and outdoor scenes.

**Collision avoidance**

It is well known that image divergence can be used in obstacle collision avoidance. Nelson and Aloimonos [163] demonstrated a robotics system which computed divergence by spatio–temporal techniques applied to the images of highly textured visible surfaces. We describe a real-time implementation based on image contours and "act" on the visually derived information.

Figure 5.5a shows a camera mounted on an Adept robot manipulator and pointing in the direction of a target contour – the lens of a pair of glasses on a mannequin. (We hope to extend this so that the robot initially searches by rotation for a contour of interest. In the present implementation, however, the target object is placed in the centre of the field of view.)

The closed contour is then localised automatically by initialising a closed loop B-spline snake in the centre of the image. The snake "explodes" outwards and deforms under the influence of image forces which cause it to be attracted to high contrast edges (figure 5.5b).

The robot manipulator then makes a deliberate motion towards the target. Tracking the area of the contour (figure 5.5c) and computing its rate of change allows us to estimate the divergence. For motion along the visual ray this is sufficient information to estimate the time to contact or impact. The estimate of time to contact – decreased by the uncertainty in the measurement and any image deformation (5.32) – is used to guide the manipulator so that it stops just before collision (figure 5.5d). The manipulator in fact, travels "blindly" after its sensing actions (above) and at a uniform speed for the time remaining until contact. In repeated trials image divergences measured at distances of 0.5m to 1.0m were estimated accurately to the nearest half of a time unit. This corresponds to a positional accuracy of 20mm for a manipulator translational velocity of 40mm/s.

The affine transformation approximation breaks down at close proximity to the target. This may lead to a degradation in the estimate of time to contact when very close to the target.

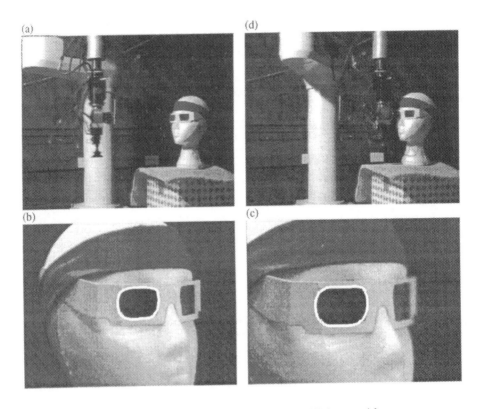

Figure 5.5: Using image divergence for collision avoidance.

*A CCD camera mounted on a robot manipulator (a) fixates on the lens of a pair of glasses worn by a mannequin (b). The contour is localised by a B-spline snake which "expands" out from a point in the centre of the image and deforms to the shape of a high contrast, closed contour (the rim of the lens). The robot then executes a deliberate motion towards the target. The image undergoes an isotropic expansion (divergence)(c) which can be estimated by tracking the closed loop snake and monitoring the rate of change of the area of the image contour. This determines the* time to contact *– a measure of the distance to the target in units of time. This is used to guide the manipulator safely to the target so that it stops before collision (d).*

Figure 5.6: Using image divergence to estimate time to contact.

*Four samples of a video sequence taken from a moving observer approaching a stationary car at a uniform velocity (approximately 1m per time unit). A B-spline snake automatically tracks the area of the rear windscreen (figure 5.7). The image divergence is used to estimate the time to contact (figure 5.8). The next image in the sequence corresponds to collision!*

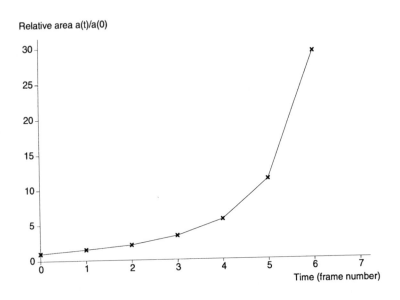

Figure 5.7: Apparent area of windscreen for approaching observer.

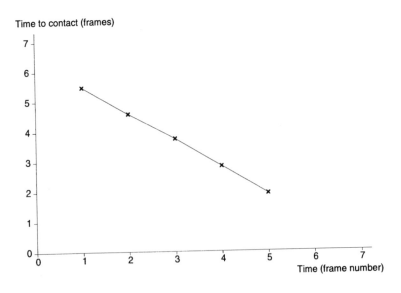

Figure 5.8: Estimated time to contact for approaching observer.

## Braking

Figure 5.6 shows a sequence of images taken by a moving observer approaching the rear windscreen of a stationary car in front. In the first frame (time $t = 0$) the relative distance between the two cars is approximately 7m. The velocity of approach is uniform and approximately 1m/time unit.

A B-spline snake is initialised in the centre of the windscreen, and expands out until it localises the closed contour of the edge of the windscreen. The snake can then automatically track the windscreen over the sequence. Figure 5.7 plots the apparent area, $a(t)$ (relative to the initial area, $a(0)$) as a function of time, $t$. For uniform translation along the optical axis the relationship between area and time is given (from (5.26) and (5.44)) by solving the first-order partial differential equation:

$$\frac{d}{dt}(a(t)) = \left(\frac{2\mathbf{U}.\mathbf{q}}{\lambda}\right) a(t). \tag{5.50}$$

Its solution is given by:

$$a(t) = \frac{a(0)}{\left[1 - \frac{t}{t^c(0)}\right]^2} \tag{5.51}$$

where $t^c(0)$ is the initial estimate of the time to contact:

$$t^c(0) = \frac{\lambda(0)}{\mathbf{U}.\mathbf{q}} \tag{5.52}$$

This is in close agreement with the data. This is more easily seen if we look at the variation of the time to contact with time. For uniform motion this should decrease linearly. The experimental results are plotted in Figure 5.8. These are obtained by dividing the area of the contour at a given time by its temporal derivative (estimated by finite differences),

$$t_c(t) = \frac{2a(t)}{a_t(t)}. \tag{5.53}$$

Their variation is linear, as predicted. These results are of useful accuracy, predicting the collision time to the nearest half time unit (corresponding to 50cm in this example).

For non-uniform motion the profile of the time to contact as a function of time is a very important cue for braking and landing reactions. Lee [133] describes experiments in which he shows that humans and animals can use this information in number of useful visual tasks. He showed that a driver must brake so that the rate of decrease of the time to contact does not exceed 0.5.

$$\frac{d}{dt}(t_c(t)) \geq -0.5. \tag{5.54}$$

The derivation of this result is straightforward. This will ensure that the vehicle can decelerate uniformly and safely to avoid a collision. As before, neither distance nor velocity appear explicitly in this expression. More surprisingly the driver needs no knowledge of the magnitude of his deceleration. Monitoring the divergence of the image velocity field affords sufficient information to control braking reactions. In the example of figure 5.6 we have shown that this can be done extremely accurately and reliably by montitoring apparent areas.

## Landing reactions and object manipulation

If the translational motion has a component parallel to the image plane, the image divergence is composed of two components. The first is the component which determines immediacy or time to contact. The other term is due to image foreshortening when the surface has a non-zero slant. The two effects can be separately computed by measuring the deformation. The deformation also allows us to recover the surface orientation.

Note that unlike stereo vision, the magnitude of the translation is not needed. Nor are the camera parameters (focal length; aspect ratio is not needed for divergence) known or calibrated. Nor are the magnitudes and directions of the camera rotations needed to keep the target in the field of view. Simple measurements of area and its moments – obtained in closed form as a function of the B-spline snake control points – were used to estimate divergence and deformation. The only assumption was of uniform motion and known direction of translation.

Figures 5.9 show two examples in which a robot manipulator uses these estimates of time to contact and surface orientation in a number of tasks including landing (approaching perpendicular to object surface) and manipulation. The tracked image contours are shown in figure 5.2. These show the effect of divergence (figure 5.2a and b) when the viewer moves towards the target, and deformation (figures 5.2c and d) due to the sideways component of translation.

## Qualitative visual navigation

Existing techniques for visual navigation have typically used stereo or the analysis of image sequences to determine the camera ego-motion and then the 3D positions of feature points. The 3D data are then analysed to determine, for example, navigable regions, obstacles or doors. An example of an alternative approach is presented. This computes qualitative information about the orientation of surfaces and times to contact from estimates of image divergence and deformation. The only requirement is that the viewer can make deliberate movements or has stereoscopic vision. Figure 5.10a shows the image of a door and

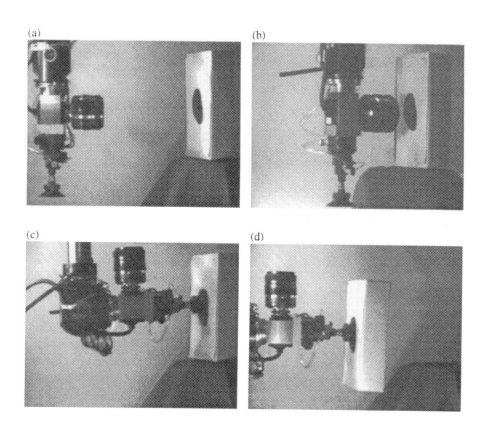

Figure 5.9: Visually guided landing and object manipulation.

*Figures 5.9 shows two examples in which a robot manipulator uses the estimates of time to contact and surface orientation in a number of tasks including landing (approaching perpendicular to object surface) and manipulation. The tracked image contours used to estimate image divergence and deformation are shown in figure 5.2.*

*In (a) and (b) the estimate of the time to contact and surface orientation is used to guide the manipulator so that it comes to rest perpendicular to the surface with a pre-determined clearance. Estimates of divergence and deformation made approximately 1m away were sufficient to estimate the target object position and orientation to the nearest 2cm in position and 1° in orientation.*

*In the second example, figures (c) and (d), this information is used to position a suction gripper in the vicinity of the surface. A contact sensor and small probing motions can then be used to refine the estimate of position and guide the suction gripper before manipulation. An accurate estimate of the surface orientation is essential. The successful execution is shown in (c) and (d).*

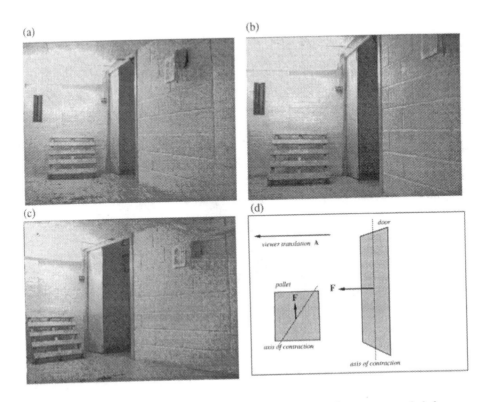

Figure 5.10: Qualitative visual navigation using image divergence and deformation.

*(a) The image of a door and an object of interest, a pallet. (b) Movement towards the door and pallet produces a deformation in the image seen as an expansion in the apparent area of the door and pallet. This can be used to determine the distance to these objects, expressed as a time to contact – the time needed for the viewer to reach the object if it continued with the same speed. (c) A movement to the left produces combinations of image deformation, divergence and rotation. This is immediately evident from both the door (positive deformation and a shear with a horizontal axis of expansion) and the pallet (clockwise rotation with shear with diagonal axis of expansion). These effects, combined with the knowledge that the movement between the images, are consistent with the door having zero tilt, i.e. horizontal direction of increasing depth, while the pallet has a tilt of approximately 90°, i.e. vertical direction of increasing depth. They are sufficient to determine the orientation of the surface qualitatively (d). This has been done with no knowledge of the intrinsic properties of the camera (camera calibration), its orientations or the translational velocities. Estimation of divergence and deformation can also be recovered by comparison of apparent areas and the orientation of edge segments.*

an object of interest, a pallet. Movement towards the door and pallet produce a deformation in the image. This is seen as an expansion in the apparent area of the door and pallet in figure 5.10b. This can be used to determine the distance to these objects, expressed as a time to contact – the time needed for the viewer to reach the object if the viewer continued with the same speed. The image deformation is not significant. Any component of deformation can, anyhow, be absorbed by (5.32) as a bound on the time to contact. A movement to the left (figure 5.10c) produces image deformation, divergence and rotation. This is immediately evident from both the door (positive deformation and a shear with a horizontal axis of expansion) and the pallet (clockwise rotation with shear with diagonal axis of expansion). These effects with the knowledge of the direction of translation between the images taken at figure 5.10a and 5.10c are consistent with the door having zero tilt, i.e. horizontal direction of increasing depth, while the pallet has a tilt of approximately $90°$, i.e. vertical direction of increasing depth. These are the effects predicted by (5.25, 5.26, 5.27 and 5.28) even though there are also strong perspective effects in the images. They are sufficient to determine the orientation of the surface qualitatively (Figure 5.10d). This has been done without knowledge of the intrinsic properties of the cameras (camera calibration), the orientations of the cameras, their rotations or translational velocities. No knowledge of epipolar geometry is used to determine exact image velocities or disparities. The solution is incomplete. It can, however, be easily augmented into a complete solution by adding additional information. Knowing the magnitude of the sideways translational velocity, for example, can determine the exact quantitative orientations of the visible surfaces.

# Chapter 6

# Conclusions

## 6.1 Summary

This thesis has presented theoretical and practical solutions to the problem of recovering reliable descriptions of curved surface shape. These have been developed from the analysis of visual motion and differential surface geometry. Emphasis has been placed on computational methods with built-in robustness to errors in the measurements and viewer motion.

It has been demonstrated that practical, efficient solutions to robotic problems using visual inferences can be obtained by:

1. Formulating visual problems in the precise language of mathematics and the methods of computation.

2. Using geometric cues such as the relative image motion of curves and the deformation of image shapes which have a resilience to and the ability to recover from errors in image measurements and viewer motion.

3. Allowing the viewer to make small, local controlled movements – active vision.

4. Taking advantage of partial, incomplete solutions which can be obtained efficiently and reliably when exact quantitative solutions are cumbersome or ill-conditioned.

These theories have been implemented and tested using a novel real-time tracking system based on B-spline snakes. The implementations of these theories are preliminary, requiring considerable effort and research to convert them into working systems.

## 6.2   Future work

The research presented in this thesis has since been extended. In conclusion we identify the directions of future work.

- Singular apparent contours
  In Chapter 2 the epipolar parameterisation was introduced as the natural parameterisation for image curves and to recover surface curvature. However the epipolar parameterisation is degenerate at singular apparent contours – the viewing ray is tangent to the contour generator (i.e an asymptotic direction of a hyperbolic surface patch) and hence the ray and contour generator do not form a basis for the tangent plane. The epipolar parameterisation can not be used to recover surface shape. Giblin and Soares [84] have shown how for orthographic projection and planar motion it is still possible to recover the surface by tracking *cusp* under known viewer motion. The geometric framework presented in Chapter 2 can be used to extended this result to arbitrary viewer motion and perspective projection.

- Structure and motion of curved surfaces
  This thesis has concentrated on the recovery of surface shape from known viewer motion. Can the deformation of apparent contours be used to solve for unknown viewer motion? This has been considered a difficult problem since each viewpoint generates a different contour generator with the contour generators "slipping" over the visible surface under viewer motion. Egomotion recovery requires a set of corresponding features visible in each view. Porril and Pollard [174] have shown how epipolar tangency points – the points on the surface where the epipolar plane is tangent to the surface – are distinct points that are visible in both views. Rieger [181] showed how in principle these points can be used to estimate viewer motion under orthographic projection and known rotation. This result can be generalised to arbitrary motion and perspective projection.

- Global descriptions of shape
  The work described in this thesis has recovered local descriptions of surface shape based on differential surface geometry. Combining these local cues and organising them into coherent groups or surfaces requires the application of more global techniques. Assembling fragments of curves and strips of surfaces into a 3D sketch must also be investigated.

- Task-directed sensor planning
  The techniques presented recover properties of a scene by looking at it from

different viewpoints. The strategies that should be used to choose the next viewpoint so as to incrementally describe the scene or accomplish a task requiring visual information must also be investigated. This requires being able to quantify the utility of exploratory movements so that they can be traded off against their cost. These ideas will be applicable to both visual navigation, path planning and grasping strategies.

- Visual attention and geometric invariants

  In the implementations presented in this thesis the object of interest has been placed in an uncluttered scene to allow the snake tracker to extract and follow the contour of interest. In less structured environments the system itself must find objects or areas of interest to which it can divert its own visual attention. We are presently investigating how to recognise target contours under arbitrary viewpoints. In principle it should be possible to recognise planar contours and to group the apparent contours of surfaces of revolution viewed under weak perspective by affine differential and integral invariants.

- Visual tracking

  Finally the B-spline snake and the availability of parallel processing hardware, e.g. the Transputer, are ripe for application to real-time visual tracking problems. The automatic initialisation of B-spline snakes on image edges, the adaptive choice of the number and position of the control points and the control of the feature search are current areas of research. B-spline snakes with reduced degrees of freedom, for example B-splines which can only deform by an affine transformation, will offer some resilience to tracking against background clutter.

# Appendix A

# Bibliographical Notes

## A.1  Stereo vision

As a consequence of the inherent ambiguity of the perspective imaging process an object feature's 3D position cannot readily be inferred from a single view. Human vision overcomes this problem in part by using multiple images of the scene obtained from different viewpoints in space. This is called stereopsis [168, 152]. In computer vision the respective paradigm is "shape from stereo" [146, 147, 152, 10, 165, 171, 8].

Binocular stereo is the most generally applicable passive method of determining the 3D structure of a scene. The basis of stereo algorithms is that the distance to a point in the scene can be computed from the relative difference in position of the projection of that point in the two images. It requires images of surfaces with sufficient texture so that distinguishing features can be identified; calibration of viewpoints; and it can only be used for features visible in both viewpoints.

The processing involved in stereopsis includes matching features that correspond to the projection of the same scene point (*the correspondence problem*); extracting disparity (the difference in image position) and then using knowledge of the camera geometry to recover the 3D structure of the objects in the scene. The most difficult part of the computation is the correspondence problem: **what to match** (pixel intensities [177], zero crossings [147, 90], edge pixels [171], line segments [8], complete high level primitives such as boundaries of closed contours [191]) and **how to match it** [147, 10, 152, 171, 165, 8].

Constraints that can be used by stereo algorithms include:

1. **Epipolar constraint:** This is a constraint derived from the imaging geometry. The line connecting the focal points of the camera systems is called the stereo baseline. Any plane containing the stereo baseline is called an epipolar plane. The intersection of an epipolar plane with an image plane is called an epipolar line. The projection of a point $P$ must lie on the

intersection of the plane that contains the stereo baseline and the point $P$ and the image planes. Every point on a given epipolar line in one image must correspond to a point on the corresponding epipolar line in the other image (figure 2.6, 3.9 and 3.10). This constraint reduces the search space from two dimensions to one.

2. **Continuity of surfaces:** Most surfaces in the real world are smooth in the sense that local variations in the surface are small compared with the overall distance from the viewer. Disparities vary smoothly except at object boundaries [146, 147, 90].

3. **Uniqueness:** Each matching primitive should match at most one primitive from the other image [146, 147].

4. **Ordering Constraint:** In a continuous surface edges must be ordered in the same way along corresponding epipolar lines unless one object lies in front of another or is transparent [10].

5. **Figural Continuity:** Edges along a contour should match edges along a similar contour in the other image, i.e. surface structure is preserved. Along matched contours, disparity should vary smoothly [152].

6. **Disparity Gradient Limit:** Nearby matches must have similar disparities while more distant matches can have a greater disparity difference [171, 176].

The correspondence problem has dominated many computational and psychophysical investigations. The correspondence problem is particularly difficult in the presence of occluding boundaries (since features are then only visible to one eye/camera) and semi-transparent surfaces such as fences or windows (since the ordering constraint may not be obeyed and a disparity gradient of two is possible along epipolar lines [101] ). Correspondence methods are often disrupted by viewpoint dependent features such as extremal boundaries or specularities (highlights) [38]. The eyes do not see the same physical feature and the stereo output will be incorrect.

Area-based (correlation) algorithms are not robust since the photometric properties of a scene are not in general invariant to viewing position as the technique assumes. Although for lambertian surfaces the shading pattern is effectively stuck to the surface and independent of view, a surface region which is tilted relative to the baseline and projected into one image will in general have a different image area to the same region projected in the other image (perspective compression of matching window or "foreshortening"). Small baseline stereo algorithms employing these techniques have however been successful with lambertian surfaces [177, 13].

Traditionally stereo has been considered as a module independent of other early vision processes/modules. The input to the stereo process may however be ambiguous and it may be preferable to use stereo models that use information from other vision processes.

Jenkin and Tsotsos [112],Waxman and Duncan [209] and Cipolla and Yamamoto [54] have attempted to unify stereo and motion analysis in a manner which helps to overcome the others shortcomings and by having motion as an additional cue in stereo matching.

Some researchers have also used knowledge about the proposed surface to resolve ambiguity in the stereo matching problem - integrating matching and surface interpolation [101, 170, 153].

## A.2   Surface reconstruction

It is widely believed that vision applications demand a dense array of depth estimates. Edge based stereo methods cannot provide this because the features are usually sparsely and irregularly distributed over the image. Psychophysical experiments with the interpretation of random-dot stereograms have shown that even with sparse data the human visual system arrives at unique interpretations of surface shape - the perception of dense and coherent surfaces in depth [113]. The visual system must invoke implicit assumptions (which reflect the general expectations about the properties of surfaces) to provide additional constraints.

Surface reconstruction is the generation of an explicit representation of a visible surface consistent with information derived from various modules (using each process as an independent source of partial information: motion and stereo for local depth and monocular cues of texture, contours and shading for local orientation). Marr [144] describes an intermediate view-centred visible surface representation – a 2 1/2 D sketch – which can be used for matching to volumetric, object centred representations. This has been pursued by a number of researchers with visual surface reconstruction being required to perform interpolation between sparse data; constraint integration and the specification of depth discontinuities (occluding contours) and orientation discontinuities (surface creases).

From sparse data, however, it is impossible to determine a unique solution surface. It is necessary to invoke additional constraints. The general approach to surface reconstruction is to trade off errors in the data to obtain the smoothest surface that passes near the data. The plausibility of possible surfaces are ranked in terms of smoothness.

Grimson [90] used the techniques of variational calculus to fit a surface which maximises a measure of "smoothness" — a regularisation term which encourages the surface to be smooth. He chose a regulariser known as the quadratic variation (formally, a Sobolev semi-norm [35]). Grimson's argument in favour of quadratic variation involves two steps. First in a Surface Consistency theory (p130, [90]) he states that for surfaces with constant albedo and isotropic reflectance the absence of zero-crossings in the laplacian of the Gaussian ($\nabla^2 G$) image means that it is unlikely that the surface has a high variation. He calls this constraint "No news is good news" because the absence of information is used as a smoothness constraint for the underlying surface. In the second step he aims to enforce this by minimising the quadratic variation in areas in which there are no zero-crossings. The derivation is not rigorous however. Minimising the surface's quadratic variation in an area is an attempt to minimise the probability of a surface inflection occurring whereas the interpretation of the surface consistency theory is that the surface has at most one surface inflection in the absence of zero-crossings [32]. The implementation of the "no news is good news" constraint is too restrictive and as with most methods whose aim is explicit surface reconstruction it suffers from over-commitment to a particular surface in the absence of sufficient information. Blake and Zisserman [31] also show that the reconstructed surface is viewpoint dependent and would "wobble" as the viewpoint is changed even though the surfaces are unchanged. This severely limits the usefulness of this and similar approaches. The success of the quadratic variation term arises from the pleasing appearance of reconstructed surfaces. This is because it is physically analogous to the energy of a thin plate.

Grimson's methods presupposed the segmentation into different surfaces since it can not localise discontinuities. Arguably the most important use of surface reconstruction from sparse data is to provide information for the segmentation process. Other researchers have extended variational techniques to localise discontinuities [192, 30]. The problem has also been formulated in terms of probabilities [83, 185]. These methods are all computationally expensive; consider deviations to be a consequence of noise only (and hence do not allow other uncertainties due to insufficient information to be represented) and suffer from over-commitment. Blake and Zisserman [31] argue that surface reconstruction methods that aim to include discontinuities suffer from the gradient limit in which smooth but slanted surfaces are considered less coherent than some surfaces with discontinuities.

An alternative approach to surface reconstruction by variational methods is to fit parameterised geometric models to the sparse depth array after clustering to find the subsets of points in three-dimensional space corresponding to significant structures in the scene. Surface models have including anything from planes via quadrics to superquadrics. All the elements in a coherent set lie on some surface in a family within some tolerance to allow for noise [57, 20, 101]. No surface model is however general enough not to break down sometimes.

## A.3   Structure from motion

Interpreting motion in space by a monocular sensor and reconstructing the depth dimension from images of different viewpoints are fundamental features of the human visual system. This is called the kinetic depth effect [206] or kineopsis [160]. In computer vision the respective paradigm is shape from monocular motion or structure from motion [201, 138].

Monocular image sequence analysis to determine motion and structure is based on the assumption of rigid body motion and can be broadly divided into 2 approaches: continuous and discrete. The *optic flow* approach relates image position and optic flow (2D velocity field that arises from the projection of moving objects on to image surface) to the underlying 3D structure and motion. These methods **either** require image velocities of points on the same 3D surface (continuous flow fields : for example [105],[1]) **or** accurate estimation of image velocity and its 1st and 2nd order spatial derivatives (local flow fields: for example [123, 138, 210]). Another approach for larger *discrete* motions extracts tokens from images in a sequence and matches them from image to image to recover the motion and structure of the environment (for example : [135, 159, 196, 197, 219, 72, 93, 214]).

Inherent difficulties include:

- **The temporal correspondence problem:** There is an aperture problem in obtaining optical flow locally [201], [105] and a correspondence problem in matching tokens in discrete views.

- **The speed–scale ambiguity:** It is impossible to determine 3D structure and motion in absolute terms for a monocular observer viewing unfamiliar objects. Both are only discernible up to a scale factor, i.e. the image motion due to a nearby object moving slowly cannot be distinguished from far-away object moving quickly. Thus it is only possible to compute distance dimensionless quantities such as the *time to contact* [189] or infer qualitative information, e.g. looming [202].

- **bas–relief ambiguity:** In addition to the *speed–scale* ambiguity a more subtle ambiguity arises when perspective effects in images are small. The bas–relief ambiguity concerns the difficulty of distinguishing between a "shallow" structure close to the viewer and "deep" structures further away [95, 94]. Note that this concerns surface orientation and its effect, unlike the speed–scale ambiguity (which leaves the structure undistorted), is to distort the structure.

- **Dealing with multiple moving objects:** This require segmentation of images into objects with the same rigid body motion [1].

- **Assumption of rigidity:** Most theories cannot cope with multiple independent motion; with non-rigidity (and hence extremal boundaries or specularities) and large amounts of noise [194]. Unfortunately the output of most existing algorithms does not degrade gracefully when these assumptions are not fully met.

Existing methods perform poorly with respect to **accuracy, sensitivity to noise,** and **robustness** in the face of errors. This is because it is difficult to estimate optic flow accurately [15], or extract the position of feature points such as corners in the image [164, 93]. Features cannot lie on a plane (or on a quadric through the two viewpoints) since these configurations of points lead to a degenerate system of equations [136, 150].

# A.4 Measurement and analysis of visual motion

The computation of visual motion can be carried out at the level of points, edges, regions or whole objects. Three basic approaches have been developed based on difference techniques, spatio-temporal gradient analysis and matching of tokens or features from image to image.

## A.4.1 Difference techniques

In many visual tracking problems not all the image is changing. It is often desirable to eliminate the stationary components and focus attention on the areas of the image which are changing. The most obvious and efficient approach is to look at difference images in which one image is subtracted from the other (pixel by pixel, or in groups of pixels) and the result is thresholded to indicate significant changes. Clusters of points with differences above threshold are assumed to correspond to portions of moving surfaces [111].

Although difference techniques are good at detecting temporal change they do not produce good velocity estimates. They are also only useful when dealing with moving objects with a stationary observer. They are however extremely fast and easy to implement.

## A.4.2 Spatio-temporal gradient techniques

These methods are based on the relation between the spatial and temporal gradients of intensity at a given point. The **spatio-temporal gradient** approach aims to estimate the motion of each pixel from one frame to the next based on the fact that for a moving object the image spatial and temporal changes are related. Consider the $xy$ plane as the image plane of the camera and the $z$ axis be the optical axis. The image sequence can be considered a time function with the intensity of a point $(x, y)$ at time $t$ given by $I(x, y, t)$. If the intensity distribution near $(x, y)$ is approximated by a plane with gradients $(I_x, I_y)$ and if this distribution is translated by $u$ in the $x$-direction and $v$ in the $y$-direction then

$$I_x u + I_y v + I_t = 0 \qquad (A.1)$$

where $I_x, I_y, I_t$ are the spatial and temporal gradients

This is the Motion Constraint equation [45, 77, 105]. By measuring the spatial and temporal gradients at a point it is possible to obtain a constraint on the image velocity of the point – namely it is possible to to compute the component of velocity in the direction of the spatial gradient. This equation assumes that the temporal change at a pixel is due to a translation of the intensity pattern. This model is only an approximation. The apparent motion of an intensity pattern may not be equivalent to the motion of 3D scene points projected into the image plane. These assumptions are usually satisfied at strong gradients of image intensity (the edges) in images and hence image velocity can be computed at edge points. It is only possible locally to determine the component of velocity perpendicular to the orientation of the edge. This is called the aperture Problem.

Spatio-temporal methods have had widespread use in visual motion measurement and applications because they do not involve an explicit correspondence stage – deciding what to match and how to match it. These methods have also been implemented using special purpose vision hardware – the Datacube [11, 39, 162] and even with purpose built VLSI chips [108].

## A.4.3 Token matching

Token matching techniques establish the correspondence of spatial image features – tokens – across frames of an image sequence. These methods have played a very important role in schemes to obtain 3D shape from image velocities (see below).

Tokens are usually based on local intensity structures, especially significant points (corners) or edges (lines and curves).

1. **Corner detection**

Initial attempts at corner detection attempted to characterise images by smooth intensity functions. Corners are defined to be positions in the image in which both the magnitude of the intensity gradient is large as well as the rate of change of gradient direction. This requires the computation of second order spatial derivatives in order to be able to compute a "curvature" in a direction perpendicular to the intensity gradient. A corner is defined as a position which maximises [68, 119, 164]:

$$\frac{I_{xx}I_y^2 + I_{yy}I_x^2 - 2I_{xy}I_xI_y}{I_{xx} + I_{yy}} \tag{A.2}$$

The computation of second order differentials is however very sensitive to noise and consequently the probability of false corner detection is high. A parallel implementation of this type of corner detector has been successfully developed at Oxford [208]. The detector uses a Datacube processor and an array of 32 T800 Transputers (320 MIPS) and a Sun workstation. For accurate corner localisation the directional derivatives are computed on raw intensities and avoid smoothing. The corner detector and tracker can track up to 400 corners at 7Hz.

An alternative approach to corner detection is by "interest operators" [157, 97] The underlying assumption is that tokens are associated with the maxima of the local autocorrelation function. Moravec's corner detector (interest operator) functions by considering a local window in the image and determining the average changes of image intensity that result from shifting the window by a small amount in various directions. If the windowed patch is a corner or isolated point then all shifts will result in a large change. This was used to detect corners. The Moravec operator suffered from a number of problems, for instance that it responds to edges as well as corners.

Harris and Stephens [97] have reformulated the Moravec operator and successfully implemented a corner detector based on the local autocorrelation at each point in the image. These are expressed in terms of Gaussian smoothed first order spatial derivatives. A high value of autocorrelation indicates the inability to translate a patch of the image in arbitrary directions without experiencing a change in the underlying intensities. Hence the value of autocorrelation can be used to indicate the presence of localisable features.

2. **Time varying edge detection**

Although corners (having 2D image structure) and the intersection of edges carry a high information content (no aperture problem) the technology for detecting these features is not as advanced or accurate as that for detecting edges (with 1D image structure) in images.

Contrast edges in images are sudden intensity changes which give rise to maxima or minima in the first spatial derivative and hence a zero-crossing (passing from positive to negative) in the second derivative. Marr and Hildreth [145] proposed finding edges in images by the zero-crossings in the Laplacian of a Gaussian filtered image. An approximation to this filter can be obtained by using the difference of two Gaussians (DOG). The importance of the technique is that it proposed finding edges at different resolutions.

Canny [48] formulated finding edges in images in terms of good detection (low probability of failing to mark real edges in the presence of image noise), good localisation (marked edges should be as close as possible to true edges) and a single response to an edge. Using a precise formulation in terms of detection and localisation and the calculus of variations Canny argued that, for step edges, the optimal detector was well approximated by a convolution with a symmetric Gaussian and directional second spatial derivatives to locate edges. This is equivalent to looking for zeros of the second derivative in direction perpendicular to the edge. The width of the Gaussian filter is chosen as a compromise between good noise suppression and localisation. The Canny edge detector has found widespread use in Computer Vision. Unfortunately real-time implementations have been restricted by the "hysteresis" stage of the algorithm in which weak edges are revived if they connect with strong edges and in which edges are thinned. Convolutions have also been computationally expensive. Recently it has been possible to produce version of the Canny edge finder operating at 5Hz using a Datacube [65].

3. **Spatio-temporal filters**

If we consider the image sequence to be a *3D image* with time (frame number) as the third dimension, an edge in a frame will appear as a plane in the 3D image. The output of a **3D edge operator** will thus give the relative magnitudes of the spatial and temporal gradients Buxton and Buxton [44] developed such a spatio-temporal filter. Heeger [98] extracts image velocities from the output of a *set* of spatio-temporal filters. A bank of 24 spatio-temporal Gabor filters runs continuously in each neighbourhood of a time-varying image. Each of the filters is tuned for a different spatial frequency and range of velocities. The output of the bank of filters is interpreted by a least-squares filter in terms of a single moving edge of variable orientation. Fahle and Poggio [70] have shown that the ability of human vision to *interpolate* visual motion — that is to see jerky motion as smooth ones, as in cinema film — is well explained in terms of such filter banks. Heeger gives an impressive display of performance of the method. An image sequence (a flight through Yosemite valley!) is processed by the filter bank to compute image velocities at all points. Then the time-varying motion field is applied to the first image frame of the sequence, causing the image to "evolve" over time. The result turns out to be a good reconstruction of the image sequence. This "proves" the quality of the recovered motion field.

### 4. Cross-correlation

Cross-correlation techniques assume that portions of the image move as a whole between frames. The image distortions induced by unrestricted motion of objects in space pose difficult problems for these techniques [139, 167].

## A.4.4 Kalman filtering

Kalman filtering is a statistical approach to linearly and recursively estimating a time-varying set of parameters ( a state vector) from noisy measurements . The Kalman filter relates a dynamic system model and the statistics of error in that model to a linear measurement model with measurement error. It has been applied to tracking problems in vision, mostly where the system model is trivial but the measurement model may be more complex [7, 92, 66]. For example edge segments in motion may be unbiased measures of the positions of underlying scene edges, but with a complex noise process which is compounded from simpler underlying pixel measurements. The Kalman filter maintains a dynamic compromise between absorbing new measurements as they are made and maintaining a memory of previous measurements. The next effect is that (in a simple filter) the filter has a memory of length $\lambda$ seconds so that measurements $t$ seconds previous are given a negatively exponential $\exp(-t/\lambda)$ weighting. The memory-length parameter $\lambda$ is essentially the ratio of measurement noise to system noise. In practice, in a multivariate filter, the memory mechanism is more complex, with many exponential time constants running simultaneously.

The full potential of the Kalman filter is not as yet exploited in vision. There is an opportunity to include non-trivial system models, as is common in other applications. For example if a moving body is being tracked, any knowledge of the possible motions of that body should be incorporated in the system model. The ballistics of a thrown projectile can be expressed as a multivariate linear differential equation. Similarly planar and linear motions can be expressed as plant models.

## A.4.5  Detection of independent motion

The ability to detect moving objects is universal in animate systems because moving objects require special attention — predator or prey? Detecting movement is also important in robotic systems. Autonomous vehicles must avoid objects that wander into their paths and surveillance systems must detect intruders.

For a stationary observer the simple approach (above) of difference images can be used to direct more sophisticated processing. For a moving observer the problem is more difficult since everything in the image may be undergoing apparent motion. The pattern of image velocities may also be quite complex. This problem is addressed by Nelson [162]. He presents a qualitative approach based on the fact that if the viewer motion is known the 2D image velocity of any stationary point in the scene must be constrained to lie on a 1-D locus in velocity – equivalent to the Epipolar constraint in stereo vision. The projected motion for an independently moving object is however unconstrained and is unlikely to fall on this locus. Nelson develops this constraint in the case of partial, inexact knowledge of the viewer's motion. In such cases he classifies the image velocity field as being one of a number of canonical motion fields — for example, translational image velocities due to observer translation parallel to the image plane. An independent moving object will be detected if it has image velocities with a component in the opposite direction of the dominant motion. Another example is that of image velocities expanding from the centre due to a viewer motion in the direction of gaze. Components of image velocity towards the origin would then indicate an independently moving object. These algorithms have been implemented using Datacube image processing boards. It processes $512 \times 512$ images, sub-sampled to $64 \times 64$ at 10Hz. Normal components of image velocity are computed using the spatial–temporal gradient approach. These velocities are then used characterise the image velocity field as being one of the canonical forms. The canonical form then determines a filter to detect independently moving objects.

## A.4.6  Visual attention

Animate vision systems employ gaze control systems to control the position of the head and eyes to to acquire, fixate and stabilise images. The main types of visual skills performed by the gaze controllers are:

**Saccadic motions** to shift attention quickly to a new area of interest without doing any visual processing;

**Foveal fixation** to put the target on the fovea and hence help to remove motion blur;

**Vergence** to keep both eyes fixated on an object of interest and hence reduce the disparities between the images as well as giving an estimate of the depth of the object;

**Smooth pursuit** to track an object of interest;

**Vestibulo-ocular reflex (VOR)** to stabilise the image when the head is moving by using knowledge of the head motion;

**Opto-kinetic reflex** using image velocities to stabilise the images.

A number of laboratories have attempted to duplicate these dynamic visual skills by building head–eye systems with head and gaze controllers as well as focus and zoom control of the cameras [58, 132, 39, 11].

## A.5    Monocular shape cues

### A.5.1    Shape from shading

Shape from shading is concerned with finding ways of deducing surface orientation from image intensity values [102, 109, 217, 103]. However image intensity values do not only depend on surface orientation alone but they also depend on how the surface is illuminated and on the surface reflectance function. Algorithms for reconstructing surfaces from shading information aim to reconstruct a surface which is everywhere consistent with observed image intensities. With assumptions of a known reflectance map, constant albedo and known illumination Horn [102] showed how differential equations relating image intensity to surface orientation could be solved.

The most tenuous of the necessary assumptions is that of known illumination. Mutual illumination effects (light bouncing off one surface and on to another before entering the eyes) are almost impossible to treat analytically when interested in recovering shape [79, 80].

As a consequence it is unlikely that shape from shading can be used to give *robust* quantitative information. It may still be possible however, to obtain incomplete but robust descriptions. Koenderink and van Doorn [128] have suggested that even with varying illumination local extrema in intensity "cling" to parabolic lines. They show that a fixed feature of the field of isophotes is the direction of the isophotes at the parabolic lines. Alternatively this can be expressed as the direction of the intensity gradient is invariant at the parabolic lines. This can be used to detect parabolic points [32]. This invariant relationship to solid shape is an important partial descriptor of visible surfaces. Description is incomplete or partial but robust since it is insensitive to illumination model. It is a *qualitative shape* descriptor.

## A.5.2   Interpreting line drawings

There are a number of different things that can give rise to intensity changes: shadows and other illumination effects; surface markings; discontinuities in surface orientation; and discontinuities in depth. From a single image it is very difficult to tell from which of these 4 things an edge is due. An important key to the 3D interpretation of images is to make explicit these edges and to separate their causes.

Interpreting line drawings of polyhedra is a well-researched subject, having been investigated since 1965 [183, 91]. The analyses and resulting algorithms are mathematically rigorous. Interpreting line drawings of curved surfaces, however, still remains an open problem.

The analysis of line drawings consists of 2 components: assignment of *qualitative* line and junction labels and *quantitative* methods to describe the relative depths of various points.

Geometric edges of an image can be labelled as convex, concave or occluding. All possible trihedral junctions can be catalogued and labelled as one of 12 types [107, 56]. Each line can only be assigned one label along its length. A constraint propagation algorithm [207] (similar to relaxation labelling [63]) is used to assign a set of consistent labels which may not, however, be unique if the drawing is ambiguous, e.g. necker reversal [89].

Even though line drawings may have legal labellings they may nonetheless be uninterpretable as drawings of polyhedra because they violate such conditions as *planarity* of edges of a face. Quantitative information such as the orientation of lines and planes is needed. Two approaches exist. Mackworth [140] presented a geometric approach using gradient space (image to plane orientation dual space) to derive a mutual constraint on the gradients of planes which intersect. Sugihara [190] reduced the line drawing after line labelling to a system of linear equations and inequalities and the interpretation of 3D structures is given as the solution of a linear programming problem.

With curved objects other labels can occur and unlike a line generated by an edge of a polyhedral object (which has just one label along its entire length) curved surface labels can change midway. Turner [200] catalogued these line labels and their possible transformations for a restricted set of curved surfaces and found an overwhelming number of junction types even without including surface markings and lines due to shadows and specularities. His junction catalogue was too large to make it practical.

Malik with the same [141] exclusions attempted to label lines as being discontinuities in depth (extremal boundaries (limbs) or tangent plane discontinuities) or discontinuities in orientation (convex or concave). He clearly expounded some of the geometric constraints.

As with other methods the output of line interpretation analysis is always incomplete. A major limitation is the multiplicity of solutions. The pre-processing to determine whether a line is a surface marking has not been developed. For real images with spurious and missing lines many junctions would be incorrectly classified. The algorithms however assume a 100% confidence in the junction labels. This is particularly severe for line drawings of curved surfaces which require the accurate detection of junctions. It is, for example, difficult to detect the difference between a L junction and a curvature-L junction in which one section is an extremal boundary. Their lack of robustness in the face of errors and noise and the presence of multiple/ambiguous solutions limit their use in real visual applications.

## A.5.3   Shape from contour

Shape from contour methods attempt to infer 3D surface orientation information from the 2D image contour in a single view. Shape from contour methods differ in the assumptions they make about the underlying surfaces and surface contours to allow them to recover 3D surface orientation information from a single view. These include:

- Isotropy of surface contour tangent directions [216]: the distribution of tangents in the image of an irregularly shaped planar curve can be used to determine the orientation of the plane.

- Extremal boundary [17]: If the object is smooth and its extremal boundary segmented the orientation of the surface at the extremal boundary can be determined.

- Planar skewed symmetries [114]: Skewed symmetries of the image contour are interpreted as projections of real orientated symmetries to give a one-parameter family of possible planar surface orientations.

- Compactness of surface [37]: The extremum principle expresses the preference for symmetric or compact 3D planar surfaces.

- Curved skewed symmetries and parallelism [188]: Parallel curves and the projection of lines of curvature on a surface can be used to determine surface orientation.

- Generalised cylinders [144]: the occluding contour of objects made of generalised cylinders can be used to determine the sign of curvature of the
  surface.

The disadvantages of these methods is that they are restricted to a particular
surface and surface contour and that they are required to make a commitment
on insufficient information. They do not allow for the inherent uncertainty. In
the absence of high-level model driven processes, however, it is impossible to
make such unique quantitative inferences.

### A.5.4   Shape from texture

Shape from texture methods aim to recover surface orientation from single
images of textured surfaces. This is achieved by making strong assumptions
about the surface texture. These include *homogeneity* (uniform density distribution of texture elements – *texels*) or *isotropy* (uniform distribution of orientations) [86, 216, 117, 62, 28, 142]. Density gradients or the distribution of
orientations of texture elements in the image are then interpreted as cues to
surface shape.

## A.6   Curved surfaces

### A.6.1   Aspect graph and singularity theory

For polyhedral objects it is possible to partition a view sphere of all possible
views into cells or regions/volumes with identical *aspects* [127], i.e. the topology
of the line drawing does not change. A typical/generic view is associated with
an open cell on the viewing sphere. For orthographic projection the cells are
bounded surfaces on the view sphere. For perspective projection the cells are
3D volumes of viewing space. In general, for small movements of the vantage
point the qualitative description of the view (aspect) does not change. If a
boundary is crossed the view can undergo a sudden, substantial change.

The viewing sphere can be decomposed into cells by boundary lines which
label the transitions between views. Two aspects are said to be connected by
an *event* and the set of all aspects has a structure of a connected graph. This
is called the aspect graph [127]. The boundaries are directly related to the
geometry of the polyhedral object: namely they correspond to the directions in
which planes disappear or reappear. Gigus and Canny [87] have attempted to
compute the aspect graphs of polyhedral objects.

For a general smooth object there are an infinite number of possible views.
To make the problem tractable it is necessary to restrict attention to generic

situations - situations which are stable under small deformations. This is equivalent in the context of vision to meaning views from a general viewpoint that are stable to small excursions in vantage.

A large body of mathematics — singularity theory — exists concerning the class of mappings from 2D manifolds to 2D manifolds and the singularities of the mapping [215, 88, 6] (see [64] for summary of major theories). The projection of a smooth surface is an example of such a surface to surface mapping. The singularities of the mapping for a smooth, featureless surface constitute the apparent contour or view. Whitney [215] showed that from a generic/general viewpoint these singularities can be of two types only: folds (a smooth one-dimensional sub-manifold) or cusps (isolated points).

As with the analysis of polyhedra it is possible to partition the view sphere into cells whose views are qualitatively similar. The boundaries separating the cells are the view directions along which a transition in views occurs as a result of the creation or elimination of cusps. Koenderink and others [125, 40, 46, 180] have taken the results of singularity theory to classify all possible generic transitions which can occur in the projected contour. These are either local ( swallowtail transition, beak to beak and lip) or multi-local transitions (triple point, cusp crossing, or tangent crossing). Callahan has related these transitions and the decomposition of the view sphere to the geometry of the object's surfaces — in particular the surface's parabolic and flecnodal curves.

The aspect graph represents in a concise way any visual experience an observer can obtain by looking at the object when traversing any orbit through space. This representation is, however, very difficult to compute and few attempts have been made at computing it for typical objects. It may be possible to simplify the aspect graph by introducing the effects of scale and excluding events that can not be detected.

The manner in which singularities change over time can provide constraints for determining an object's structure and conveying qualitative solid shape information. If in addition it is augmented with quantitative information such as surface curvatures it may prove to be a useful representation.

## A.6.2   Shape from specularities

Specularities (highlights or brilliant points) are reflected, distorted images of a light source obtained from surfaces with a specular component of reflectance.

Although they may disrupt stereo processing the behaviour of specularities under viewer motion contain valuable surface geometry cues. Koenderink and Van Doorn [128] elegantly expound the qualitative behaviour of specularities as the vantage point is moved. In particular they show that specularities travel freely in elliptic or hyperbolic regions and speed up near parabolic lines. They are created or annihilated in pairs at the parabolic lines and just before creation or destruction they move transversely to such lines. They travel most slowly in regions of high curvature and hence, for a given static view are most likely to be found in regions of high curvature.

Blake and Brelstaff [22, 25, 38] describe the detection of specularities and how to obtain local geometric information from two views provided the position of the light source is known. In particular, they show that the measurement of the stereo disparity of a specularity near a surface marking constrains the principal curvatures of the underlying surface to lie on a hyperbolic constraint curve. They show that the monocular appearance of a specularity can provide an additional constraint when it is elongated for then its axis is approximately the projection of a line of least curvature if the light source is compact.

Zisserman et al. [220] show that for a known viewer motion and without knowing the light source direction it is possible to disambiguate concave and convex surfaces. They also show that if the light source position is known, the continuous tracking of the image position of a specularity can be used to recover the locus of the reflecting points if the 3D position of at least one point is known. The required point can be found when the curve crosses an edge whose position is known from binocular stereo.

Specularities provide a powerful cue to surface geometry. However, they are usually sparsely distributed.

# Appendix B

# Orthographic projection and planar motion

A simplified derivation of the analysis of Giblin and Weiss [85] is presented. This shows how surface curvature can be recovered from image accelerations at an apparent contour. The analysis presented is valid for orthographic projection and planar viewer motion.

We wish to recover the planar contour $\mathbf{r}(\theta)$ from its orthographic projection $w(\theta)$ where $\theta$ is the orientation of the image plane (see figure B.1). In particular we would like to recover the distance to the contour, $\lambda$, and the curvature, $\kappa$. From elementary differential geometry [76] and figure B.1 these are given by:

$$\lambda = \mathbf{r}.\mathbf{T} \tag{B.1}$$

$$\kappa = \frac{1}{\mathbf{r}_\theta.\mathbf{T}} \tag{B.2}$$

and the orthographic projection $w$ is given by:

$$w = \mathbf{r}.\mathbf{N} \tag{B.3}$$

where $\mathbf{T}$ and $\mathbf{N}$ are the curve tangent and normal. Differentiating (B.3) with respect to $\theta$:

$$w_\theta = \mathbf{r}_\theta.\mathbf{N} + \mathbf{r}.\mathbf{N}_\theta \tag{B.4}$$

$$= \mathbf{r}.\mathbf{T} \tag{B.5}$$

$$\tag{B.6}$$

since the curve tangent $\mathbf{r}_\theta$ is of course perpendicular to the curve normal and the derivative of the normal with respect to $\theta$ is the tangent. The reconstructed curve is then given by:

$$\mathbf{r} = w\mathbf{N} + w_\theta\mathbf{T}. \tag{B.7}$$

Differentiating again with respect to $\theta$ we see that

$$\mathbf{r}_\theta = (w + w_{\theta\theta})\mathbf{T} \tag{B.8}$$

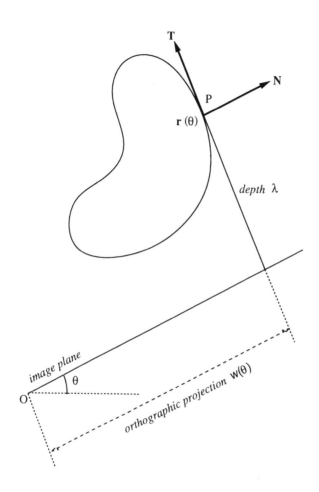

Figure B.1: A smooth contour and its orthographic projection.

and the curvature (from (B.2)) is therefore

$$\kappa = \frac{1}{(w + w_{\theta\theta})}.$$

(B.9)

Depth and curvature are obtained from first and second-order derivatives of the image with respect to viewer orientation.

# Appendix C

# Determining $\delta_{tt}.n$ from the spatio-temporal image $q(s,t)$

If the surface marking is a discrete point (image position $\mathbf{q}^*$) it is possible in principle to measure the image velocity, $\mathbf{q}_t^*$ and acceleration, $\mathbf{q}_{tt}^*$, directly from the image without any assumption about viewer motion. This is impossible for a point on an image curve. Measuring the (real) image velocity $\mathbf{q}_t$ (and acceleration $\mathbf{q}_{tt}$) for a point on an image curve requires knowledge of the viewer motion – equation (2.35). Only the normal component of image velocity can be obtained from local measurements at a curve. It is shown below however that for a discrete point–curve pair, $\boldsymbol{\delta}_{tt}.\mathbf{n}$ – the normal component of the relative image acceleration – is completely determined from measurements on the spatio-temporal image. This result is important because it demonstrates the possibility of obtaining robust inferences of surface geometry which are independent of any assumption of viewer motion.

The proof depends on re-parameterising the spatio-temporal image so that it is independent of knowledge of viewer motion. In the epipolar parameterisation of the spatio-temporal image, $\mathbf{q}(s,t)$, the $s$-parameter curves were defined to be the image contours while the $t$-parameter curves were defined by equation (2.35) so that at any instant the magnitude and direction of the tangent to a $t$-parameter curve is equal to the (real) image velocity, $\mathbf{q}_t$ – more precisely $\left.\frac{\partial \mathbf{q}}{\partial t}\right|_s$).

A parameterisation which is completely independent of knowledge of viewer motion, $\mathbf{q}(\bar{s},t)$, where $\bar{s}(s,t)$ can be chosen. Consider, for example, a parameterisation where the $t$-parameter curves (with tangent $\left.\frac{\partial \mathbf{q}}{\partial t}\right|_{\bar{s}}$) are chosen to be orthogonal to the $\bar{s}$-parameter curves (with tangent $\left.\frac{\partial \mathbf{q}}{\partial \bar{s}}\right|_t$) – the image contours. Equivalently the $t$-parameter curves are defined to be parallel to the curve normal $\mathbf{n}$,

$$\left.\frac{\partial \mathbf{q}}{\partial t}\right|_{\bar{s}} = \beta\mathbf{n} \tag{C.1}$$

where $\beta$ is the magnitude of the normal component of the (real) image velocity. Such a parameterisation can always be set up in the image. It is now possible to express the (real) image velocities and accelerations in terms of the new parameterisation.

$$q_t = \left. \frac{\partial q}{\partial t} \right|_s \tag{C.2}$$

$$= \left. \frac{\partial \bar{s}}{\partial t} \right|_s \left. \frac{\partial q}{\partial \bar{s}} \right|_t + \left. \frac{\partial q}{\partial t} \right|_{\bar{s}} \tag{C.3}$$

$$q_{tt} = \left. \frac{\partial^2 q}{\partial^2 t} \right|_s \tag{C.4}$$

$$= \left. \frac{\partial^2 \bar{s}}{\partial^2 t} \right|_s \left. \frac{\partial q}{\partial \bar{s}} \right|_t + \left( \left. \frac{\partial \bar{s}}{\partial t} \right|_s \right)^2 \left. \frac{\partial^2 q}{\partial^2 \bar{s}} \right|_t + 2 \left. \frac{\partial \bar{s}}{\partial t} \right|_s \left. \frac{\partial}{\partial \bar{s}} \left( \left. \frac{\partial q}{\partial t} \right|_{\bar{s}} \right) \right|_t + \left. \frac{\partial^2 q}{\partial^2 t} \right|_{\bar{s}}$$

$$q_{tt}.n = \left( \left. \frac{\partial \bar{s}}{\partial t} \right|_s \right)^2 \left. \frac{\partial^2 q}{\partial^2 \bar{s}} \right|_t .n + 2 \left. \frac{\partial \bar{s}}{\partial t} \right|_s \left. \frac{\partial}{\partial \bar{s}} \left( \left. \frac{\partial q}{\partial t} \right|_{\bar{s}} \right) \right|_t .n + \left. \frac{\partial^2 q}{\partial^2 t} \right|_{\bar{s}} .n. \tag{C.5}$$

From (C.3) we see that $\left( \left. \frac{\partial \bar{s}}{\partial t} \right|_s \right)$ determines the magnitude of the tangential component of image curve velocity and is not directly available from the spatio-temporal image. The other quantities in the right-hand side of the (C.5) are directly measurable from the spatio-temporal image. They are determined by the curvature of the image contour, the variation of the normal component of image velocity along the contour and the variation of the normal component of image velocity perpendicular to the image contour respectively.

However the discrete point (with image position $q^*$) which is instantaneously aligned with the extremal boundary has the same image velocity, $q_t^*$, as the point on the apparent contour. From (2.35):

$$q = q^* \tag{C.6}$$

$$q_t = q_t^*. \tag{C.7}$$

Since $q_t^*$ is measurable it allows us to determine the tangential component of the image velocity

$$\left. \frac{\partial \bar{s}}{\partial t} \right|_s = \frac{q_t . \left. \frac{\partial q}{\partial \bar{s}} \right|_t}{\left| \left. \frac{\partial q}{\partial \bar{s}} \right|_t \right|^2} \tag{C.8}$$

and hence $q_{tt}.n$ and $\delta_{tt}.n$ from spatio-temporal image measurements.

# Appendix D

# Correction for parallax based measurements when image points are not coincident

The theory relating relative inverse curvatures to the rate of parallax assumed that the two points $\mathbf{q}^{(1)}$ and $\mathbf{q}^{(2)}$ were actually coincident in the image, and that the underlying surface points were also coincident and hence at the same depth $\lambda^{(1)} = \lambda^{(2)}$. In practice, point pairs used as features will not coincide exactly. We analyse below the effects of a finite separation in image positions $\Delta\mathbf{q}$, and a difference in depths of the 2 features, $\Delta\lambda$.

$$
\begin{aligned}
\mathbf{q}^{(2)} &= \mathbf{q} \\
\mathbf{q}^{(1)} &= \mathbf{q} + \Delta\mathbf{q} \\
\lambda^{(2)} &= \lambda \\
\lambda^{(1)} &= \lambda + \Delta\lambda \\
\mathbf{q}^{(2)}.\mathbf{n} &= 0 \\
\mathbf{q}^{(1)}.\mathbf{n} &= \Delta\mathbf{q}.\mathbf{n}
\end{aligned}
\tag{D.1}
$$

If the relative inverse curvature is computed from (2.59) ,

$$
\Delta R = \frac{(\mathbf{U}.\mathbf{n})^2}{\lambda^3} \frac{1}{\boldsymbol{\delta}_{tt}.\mathbf{n}},
\tag{D.2}
$$

an error is introduced into the estimate of surface curvature due to the fact that the features are not instantaneously aligned nor at the same depth nor in the same tangent plane.

$$
R^{(2)} - R^{(1)} = \Delta R + R^{error}
\tag{D.3}
$$

where $R^{error}$ consists of errors due to the 3 effects mentioned above.

$$
R^{error} = R^{\Delta\lambda} + R^{\Delta\mathbf{q}} + R^{\mathbf{n}}
\tag{D.4}
$$

These are easily computed by looking at the differences of equation (2.56) for the 2 points. Only first-order errors are listed.

$$R^{\Delta\lambda} = \Delta\lambda\left[\frac{3}{\lambda\kappa^{t2}} + \frac{4\mathbf{U}.\mathbf{q}}{\mathbf{U}.\mathbf{n}} + \frac{\lambda\mathbf{U}_t.\mathbf{n}}{(\mathbf{U}.\mathbf{n})^2}\right]$$
$$+\Delta\lambda\left[\frac{2\lambda(\mathbf{U}.\mathbf{q})(\boldsymbol{\Omega}\wedge\mathbf{q}).\mathbf{n}}{(\mathbf{U}.\mathbf{n})^2} + \frac{\lambda(\mathbf{U}\wedge\boldsymbol{\Omega}).\mathbf{n}}{(\mathbf{U}.\mathbf{n})^2}\right] \quad\text{(D.5)}$$

$$R^{\Delta\mathbf{q}} = \frac{-2\lambda(\mathbf{U}.\boldsymbol{\delta})}{\mathbf{U}.\mathbf{n}} - \frac{\lambda^3(\boldsymbol{\Omega}_t\wedge\boldsymbol{\delta}).\mathbf{n}}{(\mathbf{U}.\mathbf{n})^2} - \frac{2\lambda^2(\mathbf{U}.\boldsymbol{\delta})(\boldsymbol{\Omega}\wedge\mathbf{q}).\mathbf{n}}{(\mathbf{U}.\mathbf{n})^2}$$
$$-\frac{2\lambda^2(\mathbf{U}.\mathbf{q})(\boldsymbol{\Omega}\wedge\boldsymbol{\delta}).\mathbf{n}}{(\mathbf{U}.\mathbf{n})^2} + \frac{\lambda^3(\boldsymbol{\Omega}.\boldsymbol{\delta})(\boldsymbol{\Omega}.\mathbf{n})}{(\mathbf{U}.\mathbf{n})^2} \quad\text{(D.6)}$$

$$R^{\mathbf{n}} = \boldsymbol{\delta}.\mathbf{n}\left[\frac{\lambda^2\mathbf{U}_t.\mathbf{q}}{(\mathbf{U}.\mathbf{n})^2} - \frac{\lambda|\mathbf{U}|^2}{(\mathbf{U}.\mathbf{n})^2} + \frac{(\mathbf{U}.\mathbf{q})}{\mathbf{U}.\mathbf{n}}\frac{1}{\kappa^{t2}} + \frac{2\lambda(\mathbf{U}.\mathbf{q})^2}{(\mathbf{U}.\mathbf{n})^2}\right]$$
$$-\boldsymbol{\delta}.\mathbf{n}\left[\frac{\lambda^2(\boldsymbol{\Omega}\wedge\mathbf{q}).\mathbf{U}}{(\mathbf{U}.\mathbf{n})^2} + \frac{\lambda^3|\boldsymbol{\Omega}|^2}{(\mathbf{U}.\mathbf{n})^2}\right] \quad\text{(D.7)}$$

# Bibliography

[1] G. Adiv. Determining three-dimensional motion and structure from optical flow generated by several moving objects. *IEEE Trans. on Pattern Analysis and Machine Intelligence*, 7(4):384–401, 1985.

[2] J.Y. Aloimonos, I. Weiss, and A. Bandyopadhyay. Active vision. In *Proc. 1st Int. Conf. on Computer Vision*, pages 35–54, 1987.

[3] A.A. Amini, S. Tehrani, and T.E. Weymouth. Using dynamic programming for minimizing the energy of active contours in the presence of hard constraints. In *Proc. 2nd Int. Conf. on Computer Vision*, pages 95–99, 1988.

[4] E. Arbogast. *Modélisation automatique d'objets non polyédriques par observation monoculaire*. PhD thesis, Institut National Polytechnique de Grenoble, 1991.

[5] R. Aris. *Vectors, Tensors, and the basic equations of fluid mechanics*. Prentice-Hall, Englewood, N.J., 1962.

[6] V. Arnold, S. Gusein-Zade, and A. Varchenko. *Singularities of Differential Maps*, volume I. Birkhauser, Boston, 1985.

[7] N. Ayache and O.D. Faugeras. Building, registration and fusing noisy visual maps. In *Proc. 1st Int. Conf. on Computer Vision*, London, 1987.

[8] N. Ayache and B. Faverjon. Efficient registration of stereo images by matching graph descriptions of edge segments. *Int. Journal of Computer Vision*, pages 107–131, 1987.

[9] R. Bajcsy. Active perception versus passive perception. In *Third IEEE Workshop on computer vision*, pages 55–59, 1985.

[10] H.H. Baker and T.O. Binford. Depth from edge and intensity based stereo. In *IJCAI*, pages 631–636, 1981.

[11] D.H. Ballard and A. Ozcandarli. Eye fixation and early vision: kinetic depth. In *Proc. 2nd Int. Conf. on Computer Vision*, pages 524–531, 1988.

[12] Y. Bar-Shalom and T.E. Fortmann. *Tracking and Data Association*. Academic Press, 1988.

[13] S.T. Barnard. A stochastic approach to stereo vision. In *Proc. 5th National Conf. on AI*, pages 676–680, 1986.

[14] S.T. Barnard and M.A. Fischler. Computational stereo. *ACM Computing Surveys*, 14(4):553–572, 1982.

[15] J. Barron. A survey of approaches for determining optic flow, environmental layout and egomotion. Technical Report RBCV-TR-84-5, University of Toronto, 1984.

[16] H.G. Barrow and J.M. Tenenbaum. Recovering intrinsic scene characteristics from images. In A. Hanson and E. Riseman, editors, *Computer Vision Systems*. Academic Press, New York, 1978.

[17] H.G. Barrow and J.M. Tenenbaum. Interpreting line drawings as three-dimensional surfaces. *Artificial Intelligence*, 17:75–116, 1981.

[18] R.H. Bartels, J.C. Beatty, and B.A. Barsky. *An Introduction to Splines for use in Computer Graphics and Geometric Modeling*. Morgan Kaufmann, 1987.

[19] F. Bergholm. Motion from flow along contours: a note on robustness and ambiguous case. *Int. Journal of Computer Vision*, 3:395–415, 1989.

[20] P.J. Besl and R.C. Jain. Segmentation through variable-order surface fitting. *IEEE Trans. Pattern Analysis and Machine Intell.*, 10(2):167–192, March 1982.

[21] L.M.H. Beusmans, D.D. Hoffman, and B.M. Bennett. Description of solid shape and its inference from occluding contours. *J. Opt. Soc. America*, 4:1155–1167, 1987.

[22] A. Blake. Specular stereo. In *IJCAI*, pages 973–976, 1985.

[23] A. Blake. Ambiguity of shading and stereo contour. In *Proc. Alvey Vision Conference*, pages 97–105, 1987.

[24] A. Blake, J.M. Brady, R. Cipolla, Z. Xie, and A. Zisserman. Visual navigation around curved obstacles. In *Proc. IEEE Int. Conf. Robotics and Automation*, volume 3, pages 2490–2495, 1991.

[25] A. Blake and G. Brelstaff. Geometry from specularities. In *Proc. 2nd Int. Conf. on Computer Vision*, pages 394–403, 1988.

[26] A. Blake and H. Bulthoff. Shape from specularities: Computation and psychophysics. *Phil. Trans. Royal Soc. London*, 331:237–252, 1991.

[27] A. Blake and R. Cipolla. Robust estimation of surface curvature from deformation of apparent contours. In O. Faugeras, editor, *Proc. 1st European Conference on Computer Vision*, pages 465–474. Springer–Verlag, 1990.

[28] A. Blake and C. Marinos. Shape from texture: estimation, isotropy and moments. *Artificial Intelligence*, 45:323–380, 1990.

[29] A. Blake, D.H. McCowen, H.R. Lo, and D. Konash. Trinocular active range-sensing. In *Proc. 1st British Machine Vision Conference*, pages 19–24, 1990.

[30] A. Blake and A. Zisserman. Some properties of weak continuity constraints and the gnc algorithms. In *Proceedings CVPR Miami*, pages 656–661, 1986.

[31] A. Blake and A. Zisserman. *Visual Reconstruction*. MIT Press, Cambridge, USA, August 1987.

[32] A. Blake, A. Zisserman, and G. Knowles. Surface description from stereo and shading. *Image and Vision Computing*, 3(4):183–191, 1985.

[33] J.D. Boissonat. Representing solids with the delaunay triangulation. In *Proc. ICPR*, pages 745–748, 1984.

[34] R.C. Bolles, H.H. Baker, and D.H. Marimont. Epipolar-plane image analysis: An approach to determining structure. *International Journal of Computer Vision*, vol.1:7–55, 1987.

[35] T.E Boult. *Information based complexity in non-linear equations and computer vision*. PhD thesis, University of Columbia, 1986.

[36] M. Brady, J. Ponce, A. Yuille, and H. Asada. Describing surfaces. *Computer Vision, Graphics and Image Processing*, 32:1–28, 1985.

[37] M. Brady and A. Yuille. An extremum principle for shape from contour. *IEEE Trans. Pattern Analysis and Machine Intell.*, 6:288–301, 1984.

[38] G. Brelstaff. *Inferring Surface Shape from Specular Reflections*. PhD thesis, Edinburgh University, 1989.

[39] C. Brown. Gaze controls cooperating through prediction. *Image and Vision Computing*, 8(1):10–17, 1990.

[40] J.W. Bruce. Seeing - the mathematical viewpoint. *The Mathematical Intelligencer*, 6(4):18–25, 1984.

[41] J.W. Bruce and P.J. Giblin. *Curves and Singularities*. Cambridge University Press, 1984.

[42] J.W. Bruce and P.J. Giblin. Outlines and their duals. In *Proc. London Math Society*, pages 552–570, 1985.

[43] T. Buchanan. 3D reconstruction from 2D images. (Unpublished seminar notes), 1987.

[44] B.F. Buxton and H. Buxton. Monocular depth perception from optical flow by space time signal processing. *Proc. Royal Society of London*, B 218:27–47, 1983.

[45] C. Cafforio and F. Rocca. Methods for measuring small displacements of television images. *IEEE Trans. on Information Theory*, vol.IT-22,no.5:573–579, 1976.

[46] J. Callahan and R. Weiss. A model for describing surface shape. In *Proc. Conf. Computer Vision and Pattern Recognition*, pages 240–245, 1985.

[47] M. Campani and A. Verri. Computing optical flow from an overconstrained system of linear algebraic equations. In *Proc. 3rd Int. Conf. on Computer Vision*, pages 22–26, 1990.

[48] J.F. Canny. A computational approach to edge detection. *IEEE Trans. Pattern Analysis and Machine Intell.*, 8:679–698, 1986.

[49] D. Charnley and R.J. Blissett. Surface reconstruction from outdoor image sequences. In *4th Alvey Vision Conference*, pages 153–158, 1988.

[50] R. Cipolla and A. Blake. The dynamic analysis of apparent contours. In *Proc. 3rd Int. Conf. on Computer Vision*, pages 616–623, Osaka, Japan, 1990.

[51] R. Cipolla and A. Blake. Surface orientation and time to contact from image divergence and deformation. In G. Sandini, editor, *Proc. 2nd European Conference on Computer Vision*, pages 187–202. Springer–Verlag, 1992.

[52] R. Cipolla and A. Blake. Surface shape from the deformation of apparent contours. *Int. Journal of Computer Vision*, 9(2):83–112, 1992.

[53] R. Cipolla and P. Kovesi. Determining object surface orientation and time
to impact from image divergence and deformation. (University of Oxford
(Memo)), 1991.

[54] R. Cipolla and M. Yamamoto. Stereoscopic tracking of bodies in motion.
*Image and Vision Computing*, 8(1):85–90, 1990.

[55] R. Cipolla and A. Zisserman. Qualitative surface shape from deformation
of image curves. *Int. Journal of Computer Vision*, 8(1):53–69, 1992.

[56] M.B. Clowes. On seeing things. *Artificial Intelligence*, 2:79–116, 1971.

[57] F.S. Cohen and D.B. Cooper. Simple parallel hierarchical and relaxation algorithms for segmenting noncausal markovian random fields. *IEEE
Trans. Pattern Analysis and Machine Intell.*, 9(2):195–219, March 1987.

[58] K.H. Cornog. Smooth pursuit and fixation for robot vision. Master's
thesis, Dept. of Elec. Eng. and Computer Science, MIT, 1985.

[59] R.M. Curwen, A. Blake, and R. Cipolla. Parallel implementation of la-
grangian dynamics for real-time snakes. In *Proc. British Machine Vision
Conference*, pages 29–35, 1991.

[60] K. Daniilidis and H-H. Nagel. Analytical results on error sensitivity of
motion estimation from two views. *Image and Vision Computing*, 8(4):297–
303, 1990.

[61] H.F. Davis and A.D. Snider. *Introduction to vector analysis*. Allyn and
Bacon, 1979.

[62] L.S. Davis, L. Janos, and S.M. Dunn. Efficient recovery of shape from
texture. *IEEE Trans. Pattern Analysis and Machine Intell.*, 5(5):485–492,
1983.

[63] L.S. Davis and A. Rosenfeld. Cooperating processes for low-level vision:
A survey. *Artificial Intelligence*, 17:47–73, 1981.

[64] M. Demazure. *Catastrophes et bifurcations*. Ecole Polytechnique, 1987.

[65] R. Deriche. Using Canny's criteria to derive a recursively implemented
optimal edge detector. *Int. Journal of Computer Vision*, 1:167–187, 1987.

[66] R. Deriche and O. Faugeras. Tracking line segments. In O. Faugeras,
editor, *Proc. 1st European Conference on Computer Vision*, pages 259–
268. Springer–Verlag, 1990.

[67] M.P. DoCarmo. *Differential Geometry of Curves and Surfaces*. Prentice-Hall, 1976.

[68] L. Dreschler and H.H. Nagel. Volumetric model and 3 trajectory of a moving car derived from monocular TV-frame sequence of a street scene. In *IJCAI*, pages 692–697, 1981.

[69] H.F. Durrant-Whyte. *Integration, Coordination, and Control of Multi-Sensor Robot Systems*. Kluwer Academic Press, Boston, MA., 1987.

[70] M. Fahle and T. Poggio. Visual hyperacuity: spatiotemporal interpolation in human vision. In *Image Understanding*, pages 49–77. Ablex, Norwood, USA, 1984.

[71] O.D. Faugeras. On the motion of 3D curves and its relationship to optical flow. In *Proceedings of 1st European Conference on Computer Vision*, 1990.

[72] O.D. Faugeras, F. Lustman, and G. Toscani. Motion and structure from motion from point and line matches. In *Proc. 1st Int. Conf. on Computer Vision*, pages 25–34, 1987.

[73] O.D. Faugeras and S.J. Maybank. Motion from point matches: multiplicity of solutions. In *IEEE Workshop on Motion*, pages 248–255, Irvine, CA., 1989.

[74] O.D. Faugeras and T. Papadopoulo. A theory of the motion fields of curves. *Int. Journal of Computer Vision*, 10(2):125–156, 1993.

[75] O.D. Faugeras and G. Toscani. The calibration problem for stereo. In *Proc. Conf. Computer Vision and Pattern Recognition*, pages 15–20, 1986.

[76] I.D. Faux and M.J. Pratt. *Computational Geometry for Design and Manufacture*. Ellis-Horwood, 1979.

[77] C.L. Fennema and W.B. Thompson. Velocity determination in scenes containing several moving objects. *Computer Vision, Graphics and Image Processing*, vol.9:301–315, 1979.

[78] M. Fleck. Some defects in finite-difference edge finders. *IEEE Trans. Pattern Analysis and Machine Intell.*, (to appear), 1991.

[79] D. Forsyth and A. Zisserman. Mutual illumination. In *Proc. Conf. Computer Vision and Pattern Recognition*, 1989.

[80] D. Forsyth and A. Zisserman. Reflections on shading. *IEEE Trans. Pattern Analysis and Machine Intell.*, 13(7):671–679, 1991.

[81] E. Francois and P. Bouthemy. Derivation of qualitative information in motion analysis. *Image and Vision Computing*, 8(4):279–288, 1990.

[82] S. Ganapathy. Decomposition of transformation matrices for robot vision. In *Proc. of IEEE Conference on Robotics*, pages 130–139, 1984.

[83] S. Geeman and D. Geeman. Stochastic relaxation, gibbs distribution and the bayesian restoration of images. *IEEE Trans. Pattern Analysis and Machine Intell.*, 6:721–741, 1984.

[84] P.J. Giblin and M.G. Soares. On the geometry of a surface and its singular profiles. *Image and Vision Computing*, 6(4):225–234, 1988.

[85] P.J. Giblin and R. Weiss. Reconstruction of surfaces from profiles. In *Proc. 1st Int. Conf. on Computer Vision*, pages 136–144, London, 1987.

[86] J.J. Gibson. *The Ecological Approach to Visual Perception*. Houghton Mifflin, 1979.

[87] Z. Gigus, J. Canny, and R. Seidel. Efficiently computing and representing aspect graphs of polyhedral objects. In *Proc. 2nd Int. Conf. on Computer Vision*, pages 30–39, 1988.

[88] M. Golubitsky and V. Guillemin. *Stable mappings and their singularities*. Springer-Verlag, 1983.

[89] R.L. Gregory. *The Intelligent Eye*. World University,London, 1970.

[90] W.E.L. Grimson. *From Images to Surfaces*. MIT Press, Cambridge, USA, 1981.

[91] A. Guzman. Computer recognition of three-dimensional objects in a scene. Technical Report MAC-TR-59, MIT, 1968.

[92] J. Hallam. Resolving obersver motion by object tracking. In *Procs. of 8th International Joint Conference on Artificial Intelligence*, volume 2, pages 792–798, 1983.

[93] C.G. Harris. Determination of ego-motion from matched points. In *3rd Alvey Vision Conference*, pages 189–192, 1987.

[94] C.G. Harris. Resolution of the bas-relief ambiguity in structure from motion under orthographic projection. In *Proc. 1st British Machine Vision Conference*, pages 67–72, 1990.

[95] C.G. Harris. Structure from motion under orthographic projection. In O. Faugeras, editor, *Proc. 1st European Conference on Computer Vision*, pages 118–123. Springer–Verlag, 1990.

[96] C.G. Harris and J.M. Pike. 3D positional integration from image sequences. In *3rd Alvey Vision Conference*, pages 233–236, 1987.

[97] C.G. Harris and M. Stephens. A combined corner and edge detector. In *4th Alvey Vision Conference*, pages 147–151, 1988.

[98] D. Heeger. Optical flow from spatiotemporal filters. In *Proc. 1st Int. Conf. on Computer Vision*, pages 181–190, 1987.

[99] H. von. Helmholtz. *Treatise on Physiological Optics*. Dover (New York), 1925.

[100] E.C. Hildreth. *The measurement of visual motion*. The MIT press, Cambridge Massachusetts, 1984.

[101] W. Hoff and N. Ahuja. Extracting surfaces from stereo images: An integrated approach. In *Proc. 1st Int. Conf. on Computer Vision*, pages 284–294, 1987.

[102] B.K.P. Horn. Understanding image intensities. *Artificial Intelligence*, 8:201–231, 1977.

[103] B.K.P. Horn. *Robot Vision*. McGraw-Hill, NY, 1986.

[104] B.K.P. Horn. Closed-form solution of absolute orientation using unit quaternions. *J. Opt. Soc. America*, A4(4):629–642, 1987.

[105] B.K.P. Horn and B.G. Schunk. Determining optical flow. *Artificial Intelligence*, vol.17:185–203, 1981.

[106] B.K.P. Horn and J.E. Weldon. Computationally effecient methods for recovering translational motion. In *Proc. 1st Int. Conf. on Computer Vision*, pages 2–11, 1987.

[107] D.A. Huffman. Impossible objects as nonsense sentences. In B. Meltzer and D. Mitchie, editors, *Machine Intelligence (6)*, pages 295–324. Edinburgh University Press, 1971.

[108] J. Hutchinson, C. Koch, J. Luo, and C. Mead. Computing motion using analog and binary resistive networks. *IEEE Computer*, March:52–63, 1988.

[109] K. Ikeuchi and B.K.P. Horn. Numerical shape from shading and occluding boundaries. *Artificial Intelligence*, 17:141–184, 1981.

[110] W.H. Ittelson. *The Ames Demonstrations in Perception.* Princeton University Press., 1952.

[111] R. Jain and H.H. Nagel. On the analysis of accumulative difference pictures from image sequences of real world scenes. *IEEE Trans. Pattern Analysis and Machine Intell.*, 1(2):206–214, 1979.

[112] M. Jenkin and J.K. Tsotsos. Applying temporal constraints to the dynamic stereo problem. *Computer, Vision Graphics and Image Processing*, vol.33:16–32, 1986.

[113] B. Julesz. *Foundations of Cyclopean Perception.* University of Chicago Press, 1971.

[114] T. Kanade. Recovery of the three-dimensional shape of an object from a single view. *Artificial Intelligence*, 17:409–460, 1981.

[115] K. Kanatani. Detecting the motion of a planar surface by line and surface integrals. *Computer Vision, Graphics and Image Processing*, 29:13–22, 1985.

[116] K. Kanatani. Structure and motion from optical flow under orthographic projection. *Computer Vision, Graphics and Image Processing*, 35:181–199, 1986.

[117] K. Kanatani and T. Chou. Shape from texture: general principle. *Artificial Intelligence*, 38:1–48, 1989.

[118] M. Kass, A. Witkin, and D. Terzopoulos. Snakes:active contour models. In *Proc. 1st Int. Conf. on Computer Vision*, pages 259–268, 1987.

[119] L. Kitchen and A. Rosenfeld. Grey-level corner detection. *Pattern Recognition Letters*, 1:95–102, 1982.

[120] J.J. Koenderink. What does the occluding contour tell us about solid shape? *Perception*, 13:321–330, 1984.

[121] J.J. Koenderink. Optic flow. *Vision Research*, 26(1):161–179, 1986.

[122] J.J. Koenderink. *Solid Shape.* MIT Press, 1990.

[123] J.J. Koenderink and A.J. Van Doorn. Invariant properties of the motion parallax field due to the movement of rigid bodies relative to an observer. *Optica Acta*, 22(9):773–791, 1975.

[124] J.J. Koenderink and A.J. Van Doorn. Geometry of binocular vision and a model for stereopsis. *Biological Cybernetics*, 21:29–35, 1976.

[125] J.J. Koenderink and A.J. Van Doorn. The singularities of the visual mapping. *Biological Cybernetics*, 24:51–59, 1976.

[126] J.J. Koenderink and A.J. Van Doorn. How an ambulant observer can construct a model of the environment from the geometrical structure of the visual inflow. In G. Hauske and E. Butenandt, editors, *Kybernetik*. Oldenburg, Munchen, 1978.

[127] J.J. Koenderink and A.J. Van Doorn. The internal representation of solid shape with respect to vision. *Biological Cybernetics*, 32:211–216, 1979.

[128] J.J. Koenderink and A.J. Van Doorn. Photometric invariants related to solid shape. *Optica Acta*, 27(7):981–996, 1980.

[129] J.J. Koenderink and A.J. Van Doorn. The shape of smooth objects and the way contours end. *Perception*, 11:129–137, 1982.

[130] J.J. Koenderink and A.J. Van Doorn. Depth and shape from differential perspective in the presence of bending deformations. *J. Opt. Soc. America*, 3(2):242–249, 1986.

[131] J.J. Koenderink and A.J. van Doorn. Affine structure from motion. *J. Opt. Soc. America*, 8(2):377–385, 1991.

[132] E. Krotkov. Focusing. *Int. Journal of Computer Vision*, 1:223–237, 1987.

[133] D.N. Lee. The optic flow field: the foundation of vision. *Phil. Trans. R. Soc. Lond.*, 290, 1980.

[134] M.M. Lipschutz. *Differential Geometry*. McGraw-Hill, NY, 1969.

[135] H.C. Longuet-Higgins. A computer algorithm for reconstructing a scene from two projections. *Nature*, 293:133–135, 1981.

[136] H.C. Longuet-Higgins. The visual ambiguity of a moving plane. *Proc.R.Soc.Lond.*, B223:165–175, 1984.

[137] H.C Longuet-Higgins. *Mental Processes:Studies in Cognitive Science*. The MIT Press, 1987.

[138] H.C. Longuet-Higgins and K. Pradzny. The interpretation of a moving retinal image. *Proc. R. Soc. Lond.*, B208:385–397, 1980.

[139] B.D. Lucas and T. Kanade. An iterative image registration technique with an application to stereo vision. In *Proc. of the 7th International Joint Conference on Artificial Intelligence*, pages 674–679, 1981.

[140] A.K. Mackworth. Interpreting pictures of polyhedral scenes. *Artificial Intelligence*, 4:121–138, 1973.

[141] J. Malik. Interpreting Line Drawings of Curved Objects. *Int. Journal of Computer Vision*, 1:73–103, 1987.

[142] C. Marinos and A. Blake. Shape from texture: the homogeneity hypothesis. In *Proc. 3rd Int. Conf. on Computer Vision*, pages 350–354, 1990.

[143] D. Marr. Analysis of occluding contour. *Proc. Royal Society, London*, 197:441–475, 1977.

[144] D. Marr. *Vision*. Freeman, San Francisco, 1982.

[145] D. Marr and E. Hildreth. Theory of edge detection. *Proc. Roy. Soc. London. B.*, 207:187–217, 1980.

[146] D. Marr and T. Poggio. Cooperative computation of stereo disparity. *Science*, vol.194:283–287, 1976.

[147] D. Marr and T. Poggio. A computational theory of human stereo vision. *Proc. R. Soc. London*, 204:301–328, 1979.

[148] S. J. Maybank. Apparent area of a rigid moving body. *Image and Vision Computing*, 5(2):111–113, 1987.

[149] S.J. Maybank. The angular velocity associated with the optical flow field arising from motion through a rigid environment. *Proc. Royal Society, London*, A401:317–326, 1985.

[150] S.J. Maybank. *A theoretical study of optical flow*. PhD thesis, Birbeck College, University of London, 1987.

[151] S.J. Maybank. The projective geometry of ambiguous surfaces. *Phil. Trans. Royal Society, London*, 332(1623):1–47, 1991.

[152] J.E.W. Mayhew and J.P. Frisby. Towards a computational and psychophysical theory of stereopsis. *AI Journal*, 17:349–385, 1981.

[153] P. McLauchlan and J. Mayhew. Structure from smooth textured surfaces. Technical Report AIVRU-037, Sheffield University, 1988.

[154] G. Medioni and Y. Yasumoto. Corner detection and curve representation using curve b-splines. In *Proc. CVPR*, pages 764–769, 1986.

[155] S. Menet, P. Saint-Marc, and G. Medioni. B-snakes:implementation and application to stereo. In *Proceedings DARPA*, pages 720–726, 1990.

[156] R.S. Millman and G.D. Parker. *Elements of Differential Geometry.* Prentice-Hall, NJ, 1977.

[157] H.P. Moravec. Visual mapping by a robot rover. In *Proc. of the 6th International Joint Conference on Artificial Intelligence*, pages 598–600, 1979.

[158] D. Murray and B. Buxton. *Experiments in the machine interpretation of visual motion.* MIT Press, Cambridge, USA, 1990.

[159] H.-H. Nagel and B. Neumann. On 3D reconstruction from two perspective views. In *Proc. of the 7th International Joint Conference on Artificial Intelligence*, pages 661–663, 1981.

[160] K. Nakayama and J.M. Loomis. Optical velocity patterns, velocity sensitive neurons and space perception: A hypothesis. *Perception*, 3:63–80, 1974.

[161] N. Navab, N. Deriche, and O.D. Faugeras. Recovering 3D motion and structure from stereo and 2D token tracking cooperation. In *Proc. 3rd Int. Conf. on Computer Vision*, pages 513–516, 1990.

[162] R.C. Nelson. Qualitative detection of motion by a moving observer. In *Proc. Conf. Computer Vision and Pattern Recognition*, 1991.

[163] R.C. Nelson and J. Aloimonos. Using flow field divergence for obstacle avoidance: towards qualitative vision. In *Proc. 2nd Int. Conf. on Computer Vision*, pages 188–196, 1988.

[164] J.A. Noble. Finding Corners. *Image and Vision Computing*, 6(2):121–128, May 1988.

[165] Y. Ohta and T. Kanade. Stereo by intra- and inter-scan line search using dynamic programming. *IEEE Trans. on Pattern Analysis and Machine Intelligence*, 7(2):139–154, 1985.

[166] B. O'Neill. *Elementary Differential Geometry.* Academic Press, 1966.

[167] G.P. Otto and T.K.W. Chau. A region-growing algorithm for matching of terrain images. In *Proc. of 4th Alvey Vision Conference*, pages 123–128, 1988.

[168] G.F. Poggio and T. Poggio. The analysis of stereopsis. *Annual Review of Neuroscience*, 7:379–412, 1984.

[169] T. Poggio, V. Torre, and C. Koch. Computational vision and regularisation theory. *Nature*, 317:314–319, 1985.

[170] S.B. Pollard. *Identifying Correspondences in Binocular Stereo*. PhD thesis, University of Sheffield, 1985.

[171] S.B. Pollard, J.E.W. Mayhew, and J.P. Frisby. PMF:A Stereo Correspondence Algorithm Using A Disparity Gradient. *Perception*, 14:449–470, 1985.

[172] J. Ponce and D.J. Kriegman. On recognizing and positioning curved 3D objects from image contours. In *Proc. IEEE workshop on the interpretation of 3D scenes*, 1989.

[173] J. Porrill. Optimal combination and constraints for geometrical sensor data. *Int. J. Robotics Research*, 7(6):66–77, 1988.

[174] J. Porrill and S.B. Pollard. Curve matching and stereo calibration. *Image and Vision Computing*, 9(1):45–50, 1991.

[175] K.F. Prazdny. Egomotion and relative depth map from optical flow. *Biological Cybernetics*, vol.36:87–102, 1980.

[176] K.F. Prazdny. Detection of binocular disparities. *Biological Cybernetics*, 1985.

[177] L.H. Quam. Hierarchical warp stereo. In *DARPA Image Understanding Workshop*, pages 149–155, 1984.

[178] C.R. Rao. *Linear Statistical Inference and Its Applications*. John Wiley and Sons, New York, 1973.

[179] I.D. Reid. *Recognizing parameterized objects from range data*. PhD thesis, University of Oxford, 1991.

[180] J. Rieger. On the classification of views of piecewise smooth objects. *Image and Vision Computing*, 5:91–97, 1987.

[181] J.H. Rieger. Three dimensional motion from fixed points of a deforming profile curve. *Optics Letters*, 11:123–125, 1986.

[182] J.H. Rieger and D.L. Lawton. Processing differential image motion. *J. Opt. Soc. America*, A2(2):354–360, 1985.

[183] L.G. Roberts. Machine perception of three - dimensional solids. In J.T. Tippet, editor, *Optical and Electro-optical Information Processing*. MIT Press, 1965.

[184] G. Scott. The alternative snake – and other animals. In *Proc. 3rd Alvey Vision Conference*, pages 341–347, 1987.

[185] J.F. Silverman and D.B. Cooper. Bayesian clustering for unsupervised estimation of surface and texture models. *IEEE Trans. Pattern Analysis and Machine Intell.*, 10(4), July 1988.

[186] D.A. Sinclair, A. Blake, and D.W. Murray. Robust ego-motion estimation. In *Proc. British Machine Vision Conference*, pages 389–394, 1990.

[187] G. Sparr and L. Nielsen. Shape and mutual cross-ratios with application to exterior, interior and relative orentation. In O. Faugeras, editor, *Proc. 1st European Conference on Computer Vision*, pages 607–609. Springer–Verlag, 1990.

[188] K.A. Stevens. The visual interpretation of surface contours. *Artificial Intelligence*, 17:47–73, 1981.

[189] M. Subbarao. Bounds on time–to–collision and rotational component from first-order derivatives of image flow. *Computer Vision, Graphics and Image Processing*, 50:329–341, 1990.

[190] K. Sugihara. An algebraic approach to shape-from-image problems. *Artificial Intelligence*, 23:59–95, 1984.

[191] H. Takahashi and F. Tomita. Self-calibration of stereo cameras. In *Proc. 2nd Int. Conf. on Computer Vision*, 1988.

[192] D. Terzopoulos. Image analysis using multigrid relaxation method. *IEEE Trans. on Pattern Analysis and Machine Intelligence*, 8(2):129–139, 1986.

[193] D.W. Thompson and J.L. Mundy. Three-dimensional model matching from an unconstrained viewpoint. In *Proceedings of IEEE Conference on Robotics and Automation*, pages 208–220, 1987.

[194] J.T. Todd. Perception of structure from motion: Is projective correspondence of moving elements a necesssary condition? *Journal of Experimental Psychology*, 11(6):689–710, 1985.

[195] R.Y. Tsai. A versatile camera calibration technique for high-accuracy 3D machine vision metrology using off-the-shelf tv cameras and lenses. *IEEE Journal of Robotics and Automation*, RA-3(4):323–344, 1987.

[196] R.Y. Tsai and T.S. Huang. Three-dimensional motion parameters of a rigid planar patch 2 : singular value decomposition. *IEEE Trans. on Acoustics, Speech and Signal Processing*, ASSP-30(4):525–534, 1982.

[197] R.Y. Tsai and T.S. Huang. Uniqueness and estimation of three-dimensional motion parameters of a rigid objects with curved surfaces. *IEEE Trans. on Pattern Analysis and Machine Intelligence*, 6(1):13–26, 1984.

[198] R.Y. Tsai and R.K. Lenz. A new technique for fully autonomous and efficient 3D robotics hand-eye calibration. In *4th International Symposium on Robotics Research*, volume 4, pages 287–297, 1987.

[199] R.Y. Tsai and R.K. Lenz. Techniques for calibration of the scale factor and image center for high accuracy 3D machine vision metrology. *IEEE Trans. Pattern Analysis and Machine Intell.*, 10(5):713–720, 1988.

[200] K.J. Turner. *Computer perception of curved objects using a television camera.* PhD thesis, University of Edinburgh, 1974.

[201] S. Ullman. *The interpretation of visual motion.* MIT Press, Cambridge,USA, 1979.

[202] S. Ullman. Analysis of visual motion by biological and computer vision. *IEEE COMPUTER*, vol.14(8):57–69, 1981.

[203] R. Vaillant. Using occluding contours for 3D object modelling. In O. Faugeras, editor, *Proc. 1st European Conference on Computer Vision*, pages 454–464. Springer–Verlag, 1990.

[204] R. Vaillant and O.D. Faugeras. Using extremal boundaries for 3D object modelling. *IEEE Trans. Pattern Analysis and Machine Intell.*, 14(2):157–173, 1992.

[205] A. Verri and A Yuille. Some perspective projection invariants. *J. Opt. Soc. America*, A5(3):426–431, 1988.

[206] H. Wallach and D.N. O'Connell. The kinetic depth effect. *Journal of Experimental Psychology*, 45:205–217, 1963.

[207] D. Waltz. Understanding line drawings of scenes with shadows. In Winston. P.H., editor, *The Psychology of Vision*. McGraw-Hill,New York, 1975.

[208] H. Wang. Corner detection for 3D vision using array processors. In *Proc. BARNAIMAGE 91*, Barcelona, 1991.

[209] A.M. Waxman and J.H. Duncan. Binocular image flows : Steps toward stereo-motion fusion. *IEEE Trans. on Pattern Analysis and Machine Intelligence*, 8(6):715–729, 1986.

[210] A.M. Waxman and S. Ullman. Surface structure and three-dimensional motion from image flow kinematics. *Int. Journal of Robotics Research*, 4(3):72–94, 1985.

[211] A.M. Waxman and K. Wohn. Contour evolution, neighbourhood deformation and global image flow: planar surfaces in motion. *Int. Journal of Robotics Research*, 4(3):95–108, 1985.

[212] D. Weinshall. Direct computation of 3D shape and motion invariants. In *Proc. 3rd Int. Conf. on Computer Vision*, pages 230–237, 1990.

[213] D. Weinshall. Qualitative depth from stereo, with applications. *Computer Vision, Graphics and Image Processing*, 49:222–241, 1990.

[214] J. Weng, T.S. Huang, and N. Ahuja. Motion and structure from two perspective views: Algorithms, error analysis, and error estimation. *IEEE Trans. Pattern Analysis and Machine Intell.*, 11(5), 1989.

[215] H. Whitney. On singularities of mappings of euclidean spaces: mappings of the plane into the plane. *Ann. Math.*, 62:374–410, 1955.

[216] A.P. Witkin. Recovering surface shape and orientation from texture. *Artificial Intelligence*, 17:17–45, 1981.

[217] R.J. Woodham. Analysing images of curved surfaces. *Artificial Intelligence*, 17:117–140, 1981.

[218] M. Yamamoto. Motion analysis using the visualized locus method. *Trans. of Information Processing Society of Japan*, vol.22,no.5:442–449, 1981. (in Japanese).

[219] X. Zhuang, T.S. Huang, and R. M. Haralick. Two view motion analysis : a unified algorithm. *J. Opt. Soc. America*, 3(9):1492–1500, 1986.

[220] A. Zisserman, P.J. Giblin, and A. Blake. The information available to a moving observer from specularities. *Image and Vision Computing*, 7(1):38–42, 1989.

# Lecture Notes in Computer Science

For information about Vols. 1–949

please contact your bookseller or Springer-Verlag

Vol. 985: T. Sellis (Ed.), Rules in Database Systems. Proceedings, 1995. VIII, 373 pages. 1995.

Vol. 986: Henry G. Baker (Ed.), Memory Management. Proceedings, 1995. XII, 417 pages. 1995.

Vol. 987: P.E. Camurati, H. Eveking (Eds.), Correct Hardware Design and Verification Methods. Proceedings, 1995. VIII, 342 pages. 1995.

Vol. 988: A.U. Frank, W. Kuhn (Eds.), Spatial Information Theory. Proceedings, 1995. XIII, 571 pages. 1995.

Vol. 989: W. Schäfer, P. Botella (Eds.), Software Engineering — ESEC '95. Proceedings, 1995. XII, 519 pages. 1995.

Vol. 990: C. Pinto-Ferreira, N.J. Mamede (Eds.), Progress in Artificial Intelligence. Proceedings, 1995. XIV, 487 pages. 1995. (Subseries LNAI).

Vol. 991: J. Wainer, A. Carvalho (Eds.), Advances in Artificial Intelligence. Proceedings, 1995. XII, 342 pages. 1995. (Subseries LNAI).

Vol. 992: M. Gori, G. Soda (Eds.), Topics in Artificial Intelligence. Proceedings, 1995. XII, 451 pages. 1995. (Subseries LNAI).

Vol. 993: T.C. Fogarty (Ed.), Evolutionary Computing. Proceedings, 1995. VIII, 264 pages. 1995.

Vol. 994: M. Hebert, J. Ponce, T. Boult, A. Gross (Eds.), Object Representation in Computer Vision. Proceedings, 1994. VIII, 359 pages. 1995.

Vol. 995: S.M. Müller, W.J. Paul, The Complexity of Simple Computer Architectures. XII, 270 pages. 1995.

Vol. 996: P. Dybjer, B. Nordström, J. Smith (Eds.), Types for Proofs and Programs. Proceedings, 1994. X, 202 pages. 1995.

Vol. 997: K.P. Jantke, T. Shinohara, T. Zeugmann (Eds.), Algorithmic Learning Theory. Proceedings, 1995. XV, 319 pages. 1995.

Vol. 998: A. Clarke, M. Campolargo, N. Karatzas (Eds.), Bringing Telecommunication Services to the People – IS&N '95. Proceedings, 1995. XII, 510 pages. 1995.

Vol. 999: P. Antsaklis, W. Kohn, A. Nerode, S. Sastry (Eds.), Hybrid Systems II. VIII, 569 pages. 1995.

Vol. 1000: J. van Leeuwen (Ed.), Computer Science Today. XIV, 643 pages. 1995.

Vol. 1002: J.J. Kistler, Disconnected Operation in a Distributed File System. XIX, 249 pages. 1995.

VOL. 1003: P. Pandurang Nayak, Automated Modeling of Physical Systems. XXI, 232 pages. 1995. (Subseries LNAI).

Vol. 1004: J. Staples, P. Eades, N. Katoh, A. Moffat (Eds.), Algorithms and Computation. Proceedings, 1995. XV, 440 pages. 1995.

Vol. 1005: J. Estublier (Ed.), Software Configuration Management. Proceedings, 1995. IX, 311 pages. 1995.

Vol. 1006: S. Bhalla (Ed.), Information Systems and Data Management. Proceedings, 1995. IX, 321 pages. 1995.

Vol. 1007: A. Bosselaers, B. Preneel (Eds.), Integrity Primitives for Secure Information Systems. VII, 239 pages. 1995.

Vol. 1008: B. Preneel (Ed.), Fast Software Encryption. Proceedings, 1994. VIII, 367 pages. 1995.

Vol. 1009: M. Broy, S. Jähnichen (Eds.), KORSO: Methods, Languages, and Tools for the Construction of Correct Software. X, 449 pages. 1995. Vol.

Vol. 1010: M. Veloso, A. Aamodt (Eds.), Case-Based Reasoning Research and Development. Proceedings, 1995. X, 576 pages. 1995. (Subseries LNAI).

Vol. 1011: T. Furuhashi (Ed.), Advances in Fuzzy Logic, Neural Networks and Genetic Algorithms. Proceedings, 1994. (Subseries LNAI).

Vol. 1012: M. Bartosˇek, J. Staudek, J. Wiedermann (Eds.), SOFSEM '95: Theory and Practice of Informatics. Proceedings, 1995. XI, 499 pages. 1995.

Vol. 1013: T.W. Ling, A.O. Mendelzon, L. Vieille (Eds.), Deductive and Object-Oriented Databases. Proceedings, 1995. XIV, 557 pages. 1995.

Vol. 1014: A.P. del Pobil, M.A. Serna, Spatial Representation and Motion Planning. XII, 242 pages. 1995.

Vol. 1015: B. Blumenthal, J. Gornostaev, C. Unger (Eds.), Human-Computer Interaction. Proceedings, 1995. VIII, 203 pages. 1995.

VOL. 1016: R. Cipolla, Active Visual Inference of Surface Shape. XII, 194 pages. 1995.

Vol. 1017: M. Nagl (Ed.), Graph-Theoretic Concepts in Computer Science. Proceedings, 1995. XI, 406 pages. 1995.

Vol. 1018: T.D.C. Little, R. Gusella (Eds.), Network and Operating Systems Support for Digital Audio and Video. Proceedings, 1995. XI, 357 pages. 1995.

Vol. 1019: E. Brinksma, W.R. Cleaveland, K.G. Larsen, T. Margaria, B. Steffen (Eds.), Tools and Algorithms for the Construction and Analysis of Systems. Selected Papers, 1995. VII, 291 pages. 1995.

Vol. 1020: I.D. Watson (Ed.), Progress in Case-Based Reasoning. Proceedings, 1995. VIII, 209 pages. 1995. (Subseries LNAI).

Vol. 1021: M.P. Papazoglou (Ed.), OOER '95: Object-Oriented and Entity-Relationship Modeling. Proceedings, 1995. XVII, 451 pages. 1995.

Vol. 1022: P.H. Hartel, R. Plasmeijer (Eds.), Functional Programming Languages in Education. Proceedings, 1995. X, 309 pages. 1995.

Vol. 1023: K. Kanchanasut, J.-J. Lévy (Eds.), Algorithms, Concurrency and Knowlwdge. Proceedings, 1995. X, 410 pages. 1995.

Vol. 1024: R.T. Chin, H.H.S. Ip, A.C. Naiman, T.-C. Pong (Eds.), Image Analysis Applications and Computer Graphics. Proceedings, 1995. XVI, 533 pages. 1995.

Vol. 1025: C. Boyd (Ed.), Cryptography and Coding. Proceedings, 1995. IX, 291 pages. 1995.

Vol. 1026: P.S. Thiagarajan (Ed.), Foundations of Software Technology and Theoretical Computer Science. Proceedings, 1995. XII, 515 pages. 1995.

Vol. 1027: F.J. Brandenburg (Ed.), Graph Drawing. Proceedings, 1995. XII, 526 pages. 1996.